T0074572

Hubble Deep Field and the Distant Universe

AAS | IOP Astronomy

AAS Editor in Chief

About the program:

AAS-IOP Astronomy ebooks is the official book program of the American Astronomical Society (AAS), and aims to share in depth the most fascinating areas of astronomy, astrophysics, solar physics and planetary science. The program includes publications in the following topics:

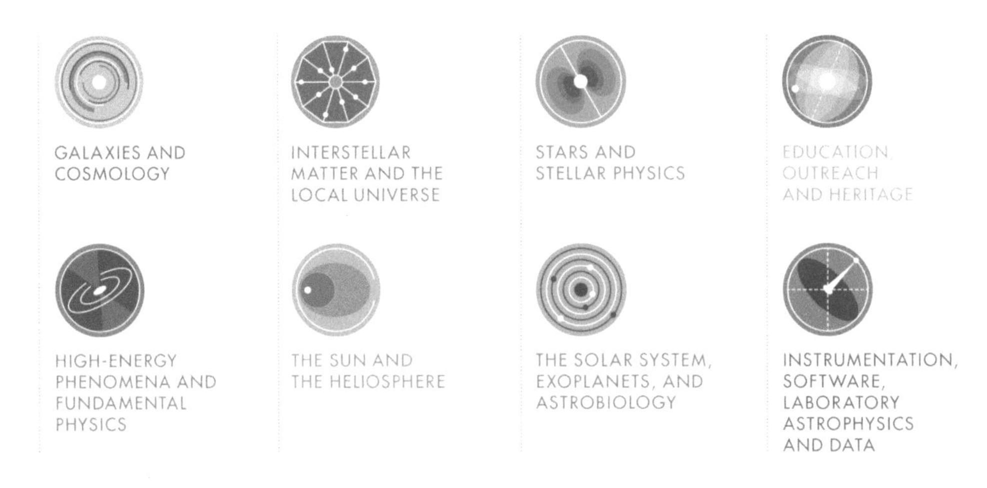

Books in the program range in level from short introductory texts on fast-moving areas, graduate and upper-level undergraduate textbooks, research monographs and practical handbooks.

For a complete list of published and forthcoming titles, please visit iopscience.org/books/aas.

About the American Astronomical Society

The American Astronomical Society (aas.org), established 1899, is the major organization of professional astronomers in North America. The membership (~7,000) also includes physicists, mathematicians, geologists, engineers and others whose research interests lie within the broad spectrum of subjects now comprising the contemporary astronomical sciences. The mission of the Society is to enhance and share humanity's scientific understanding of the universe.

Hubble Deep Field and the Distant Universe

Robert Williams
Space Telescope Science Institute, Baltimore, MD 21218, USA
University of California-Santa Cruz, Santa Cruz, CA 95064, USA

IOP Publishing, Bristol, UK

Multimedia content is available for this book from http://iopscience.iop.org/book/978-0-7503-1756-6.

ISBN 978-0-7503-1756-6 (ebook)
ISBN 978-0-7503-1754-2 (print)
ISBN 978-0-7503-1755-9 (mobi)

DOI 10.1088/978-0-7503-1756-6

Version: 20181001

AAS–IOP Astronomy
ISSN 2514-3433 (online)
ISSN 2515-141X (print)

British Library Cataloguing-in-Publication Data: A catalogue record for this book is available from the British Library.

Published by IOP Publishing, wholly owned by The Institute of Physics, London

IOP Publishing, Temple Circus, Temple Way, Bristol, BS1 6HG, UK

US Office: IOP Publishing, Inc., 190 North Independence Mall West, Suite 601, Philadelphia, PA 19106, USA

For Elaine
whose work in autism has transformed so many lives

And for my sixteen colleagues on the original HDF team
whose dedicated efforts revealed the furthest reaches of the universe

Contents

Preface

This book was born one afternoon when I received a phone call from my colleague Dr. Melissa McGrath, who informed me that she was on a committee that was interested in receiving manuscripts on astronomical topics for a new series of books underwritten by the American Astronomical Society in collaboration with the Institute of Physics/UK. Might I consider writing something at the university graduate level on the science that came out of the Hubble Deep Field? I allowed that my research interests had evolved away from the distant universe in the intervening 20 years since the HDF Project had been carried out, but I would speak with some of my colleagues who were still doing research in that area about a possible collaboration.

I felt that a less technical-level book might be of some interest to a wider readership of non-scientists, yet could serve as a good guide to what science had been learned from the HDF. I was interested in writing something at a more popular level that described the history and process by which the Deep Field observation came into being, what the major results were, and how the Project helped change the culture in which large projects on forefront telescopes might be carried out. What finally emerged in the two years following that initial phone conversation is a book that is an amalgam of what I wanted to write and what the IOP wanted to receive.

The book contains a wide range of levels that no doubt make parts of it a mismatch for some audiences. The penultimate chapter on star formation and photometric redshifts is hardly bedtime reading for most, however it should be understandable to the serious amateur astronomer who wants to understand the profound things that distant galaxies in the Deep Field are telling us. In other respects, the book should be quite understandable and informative about the large-scale structure and evolution of the universe for the lay reader.

The undertaking of the original HDF owes much to many people. My sixteen colleagues on the team who brought the HDF image to life can never be thanked enough for their devotion to a risky project that required such focus and effort. They are the true heroes and heroines. Thanks are due to my colleague Anton Koekemoer, who has been involved with many of the follow-up deep field programs on *Hubble Space Telescope,* and whose thoughts and perspective were very helpful in the organization of the book.

My deepest thanks go to Drs. Richard Ellis, Garth Illingworth, Sandra Faber, Harry Ferguson, Mark Dickinson, and Alan Dressler for always responding knowledgeably and diplomatically to my many queries in recent years about their seminal research on galaxies near and far—grand objects in this fascinating cosmos that represent the roots of not just planet Earth, but life itself.

R. E. Williams

July 2018

Author biography

Robert Williams

Robert Williams is Distinguished Osterbrock Visiting Professor at University of California/Santa Cruz, and Astronomer Emeritus at the Space Telescope Science Institute (STScI) in Baltimore, MD, having served as Director of the Institute from 1993 to 1998. The Institute, together with Goddard Space Flight Center, operates the *Hubble Space Telescope* for NASA. Before assuming his present positions Williams spent eight years in Chile as Director of the Cerro Tololo Inter-American Observatory, the national observatory of the U.S. in the southern hemisphere. Prior to that time, he was Professor of Astronomy at the University of Arizona in Tucson for 18 years. Dr. Williams' research specialties are nebulae, novae, and emission-line spectroscopy and analysis. He is a strong advocate for science education and outreach and has lectured around the world on the discoveries from *Hubble* telescope and other forefront facilities on the ground and in space.

Williams received his undergraduate degree from the University of California, Berkeley in 1962, and a PhD in astronomy from the University of Wisconsin in 1965. He was Senior Fulbright Professor at University College London from 1972 to 1973, and received the Alexander von Humboldt Award from the German government in 1991. In 1998, he was awarded the Beatrice Tinsley Prize of the American Astronomical Society for his leadership of the Hubble Deep Field Project, which revealed in remarkable detail the early universe with *Hubble Space Telescope*. For this project, he was awarded the NASA Distinguished Public Service Medal in 1999.

Dr. Williams is an elected member of the American Academy of Arts & Sciences, and is Past-President of the International Astronomical Union, a worldwide organization serving professional astronomers and promoting international astronomy. In 2016, he was awarded the Karl Schwarzschild Medal for career achievement in astrophysics by the German Astronomische Gesellschaft. He resides in Baltimore with his wife Elaine, a pediatric psychologist who specializes in the diagnosis and treatment of autism disorders; they are co-founders of a non-profit organization in Baltimore that places adults with autism in the workplace.

AAS | IOP Astronomy

Hubble Deep Field and the Distant Universe

Robert Williams

Chapter 1

The Beginnings

The Hubble Deep Field, or HDF as it is commonly called, was an image of an insignificant patch of sky that contained thousands of galaxies of different shapes, sizes, and distances that enabled astronomers to look back in time and construct the basic history of structure in the universe. The faintest galaxies in the image are so distant their light has taken billions of years to arrive at our telescope, and therefore we are able to probe the evolution of the cosmos from the epoch when galaxies first formed just after the Big Bang to the present time. Observing galaxies at different distances is akin to taking a core sample of the Earth to understand its evolution. Every layer in the core represents a distinct time, so layers stacked upon each other reveal the changing properties of the ground over time. The evolution of galaxies in the universe can be pieced together from the HDF image in the same way—galaxies at different distances represent the universe at the time in the past when that light left those galaxies on its journey toward Earth. We see those galaxies and stars directly as they were eons ago.

Several previous mosaics of the sky obtained from small satellite telescopes at longer microwave wavelengths in the past two decades had actually succeeded in probing deeper than the HDF into the primeval hazy formless universe before structure had truly formed. For example, the mosaic from the recent NASA *WMAP* satellite, shown in Figure 1.1, used the greater penetrating ability of microwaves to peer through intervening dust and haze to a time only a few hundreds of thousands of years following the Big Bang, 13.7 billion years ago. The microwave images that make up these previous mosaics were very low-definition. That is, they were not capable of resolving anything as small as the sizes of galaxies. They were only able to distinguish immensely larger, fuzzy gaseous forms in the early universe before any definite structures had yet developed. In this sense, they provided astronomers with important knowledge of the temperature and density of the diffuse gas in the universe that was created out of the energy of the Big Bang, and the fact that the Big Bang did initiate an impressive expansion of hot, formless gas that was remarkably homogeneous.

doi:10.1088/978-0-7503-1756-6ch1

Figure 1.1. An image of the distant (and therefore very early) universe shortly after the Big Bang, taken by NASA's *WMAP* satellite at microwave wavelengths. The different colors represent huge complexes of gas, thousands of times larger than galaxies. The blue–yellow colors denote minute differences in the temperature and density of the gas that filled the universe at that time. The largest temperature differences amount to only 0.0001° [credit: NASA/*WMAP* Team].

The indistinct image of radiation originating in the distant, early universe shown in Figure 1.1 represents what is called the microwave background—the fuzzy birth image of the universe that is seen in background when one looks out in all directions. The different colors in the image denote tiny differences in the density and temperature of the initial hot gas that shows the early universe to have been very uniform in nature. What is remarkable about the gas that emerged from the Big Bang creation event is that those minute differences in the density of that gas, which is far more uniform than the air we breathe in a still room, represent condensations in the expanding cosmos that were eventually collapsed by their own gravity into huge concentrations of galaxies. The fact that all structure that now exists—planets, stars, nebular gas clouds, and even life forms—developed with the earliest galaxies that formed out of that haze and are seen in the Hubble Deep Field, truly qualifies the HDF as the universe's baby (if not birth) picture. The difference between the indistinct *WMAP* microwave background picture in Figure 1.1 and a detail of the Deep Field, which is shown for comparison in Figure 1.2, is impressive.

Taken with a modern instrument attached to the technologically advanced *Hubble Space Telescope*, the HDF's success in revealing the path for fundamental studies of the universe is the culmination of centuries of effort by thoughtful individuals who have tried to understand the workings of the cosmos and whose thinking and experimenting has set the course for succeeding generations to try their best to ferret out truth from a disparate sea of facts. It was they who laid the groundwork for much that followed in our attempts to understand the universe. Had it not been for the advocacy of these individuals, who debated with their colleagues new ways to think about the universe, the HDF might have ended up being nothing more than a curious picture of strangely shaped blobs. Instead, a succession of free-thinking individuals found their understanding of nature to be so different from the conventional thinking of their time that they persisted in advocating different ways

Figure 1.2. A portion of the Hubble Deep Field image that shows galaxies at many distances. The brightest, largest galaxies are comparatively nearby, "only" a billion light years from our Milky Way galaxy. The faintest, smallest, amorphous, and barely visible galaxies are at distances of up to 10 billion light years. All galaxies formed out of the indistinct mass of gas represented in Figure 1.1 [credit: STScI/Z. Levay].

of understanding natural phenomena—even, regrettably, on pain of death. Progress came slowly but it did finally evolve from thought that was steeped in mysticism and superstition to what we now call the scientific method that made the HDF possible.

The Greek scholar Aristotle (384–322 BCE) serves as a good starting point for understanding how observations are interpreted by scientists in modern times, in that he was one of the first individuals who documented his attempts to contemplate the physical world, trying to understand in a consistent way why things were the way there were. His starting basis was that there were certain universal, philosophical truths that influenced how the physical world worked. Although we would not now consider his *truths* to have a scientific basis, Aristotle also made observations and gathered facts in order to explain phenomena. His belief that the Earth, rather than the Sun, was the center of the solar system was incorrect: rather, it was a supposed "truth" based upon a preconceived notion held by the Greeks about how the cosmos should be ordered. Nevertheless, Aristotle did arrive at his geocentric model of the solar system with the aid of observations, albeit imprecise. In his construction of the ordering of the cosmos, when the observations went counter to prevailing conceptions of the time, Aristotle did accept the results of his observations.

When Latin translations of many of Aristotle's works became available in 12th century Europe, various philosopher theologians took elements of Aristotle's methods and modified them to formulate important ingredients of the core of modern scientific thinking. Robert Grosseteste (1168–1253), Bishop of Lincoln, presented an initial treatment of knowledge, science, and intuition, and was one of the first to include the use of mathematics as a framework for understanding the physical world. He laid out the basic optical principles of a lens, depicted in Figure 1.3, demonstrating the refraction of light by water, writing that "it is clear how to design the shape of the transparent medium in order to . . . make small things placed at a distance appear any size we want." The ideas of Grosseteste were developed by a number of 13th century scholars, including the Franciscan friar Roger Bacon (1214–1292), who was critical of Aristotle's writings and emphasized the need for experimental validation of hypotheses in discovering scientific principles. Centuries later, the English statesman Sir Francis Bacon (1561–1626) expanded this into a more developed empirical method of performing experiments, which underscores science today.

One of the most important developments in the evolution of scientific thinking, advocated strongly by the Franciscan theologian William of Ockham (1285–1347), was the idea that unsupported preconceived ideas should all be subjected to strict examination. In his writings, Ockham emphasized the importance of applying the principle of frugality in attempting to explain phenomena. Known today as "Ockham's Razor," he postulated that "entities should not be multiplied beyond

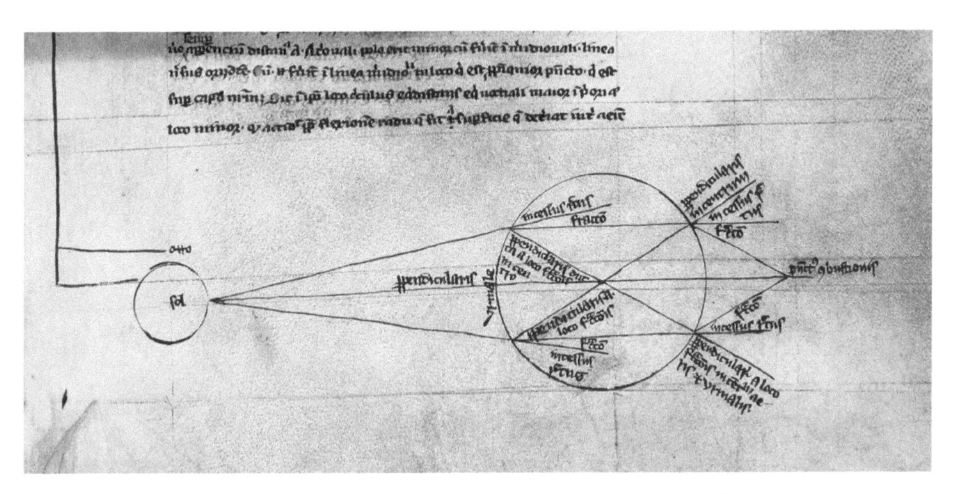

Figure 1.3. An illustration published by Roger Bacon in the 13th century showing how water refracts, or bends, light to a focus based upon Grosseteste's concepts of light and optics [credit: British Museum].

necessity." This procedure of attempting to understand phenomena in a simple way by "shaving" away unnecessary hypotheses has become a fundamental cornerstone of scientific thinking.

Hypotheses have merit within the scientific domain if and only if a justification can be given for them and they can be tested in some way. One of the most important examples of this was Copernicus' (1473–1543) advocacy of the Sun-centered planetary system, in opposition to the geocentric solar system supported by Aristotle, which had been believed firmly for many centuries. The Copernican hypothesis, or model, explaining the apparent motion of the Sun and planets was motivated by its much simpler explanation of the retrograde (i.e., apparent backward) motion of planets in the sky at various times in their orbits. The prevailing Aristotelian and Ptolemaic Earth-centered planetary system hypothesized without evidence that the orbits of planets must consist of a series of circular components in an extremely complex succession of epicycles in order to explain retrograde motion. A controversial hypothesis at the time, the much simpler Copernican system was eventually validated by observation centuries later when the necessarily precise measurements of stellar motion became feasible.

As remote as William of Ockham's philosophy and the legacy of Ockham's Razor may seem to the HDF, the scientific method owes much to his persistent debates around the European continent discrediting the validity of ad hoc hypotheses. Ockham's Razor is central to the scientific culture that has produced so many profound discoveries. It is important to understand that reducing hypotheses does not necessarily lead directly to the correct explanation of an event or situation. Rather, experience has shown that following Ockham's Razor to keep ad hoc hypotheses to a minimum is, over the long run, a much more efficient way to advance science via the elimination of false leads.

A priori, one does not know which hypotheses are false and which are valid, and all may frequently appear to have some justification. The most effective way to determine their validity is to perform experiments—or observations, in the case of astronomy—that can whittle away hypotheses or models that are not correct in order to formulate and verify those that are. It was precisely this philosophy that drove the Deep Field Project in spite of arguments that were put forth that it was a bad idea, most likely to fail. However, its success did not depend so much on the robust scientific method, as crucial as it is, as it did on innovations and a uniqueness that made the *Hubble Space Telescope* (universally referred to as the *HST* within the scientific community) capable of seeing further and fainter objects than any other telescope before it.

Advances in technology inevitably produce new discoveries by enabling observations and experiments that previously were not possible. In the case of astronomy—which is an observational science, rather than a controllable one that allows experiments to be constructed in a lab—a succession of important innovations occurred in the latter half of the past century that revolutionized the way telescopes could be built and how they could be operated. It also produced dramatically more sensitive detectors for instruments. Taken together with the development of large computers that enabled the detailed interpretation of data, these advances constituted a shaking of the foundations for observational astronomy. One could not have hoped for a better alignment of technological change with the emergence of fundamental questions that had arisen from previous advances in physics.

Early in the 20th century, Einstein's theory of general relativity and the development of quantum mechanics were totally new paradigms that were necessary to understand the behavior of nature on both enormous cosmological and minute subatomic scales. They posed fundamental questions about the nature of the universe that motivated astronomers to address them with their increasingly sophisticated telescopes. What provides the energy by which stars shine? How were the elements formed? Do planets orbit other stars? Can extremely massive objects collapse into bizarre configurations as black holes? Do they instead explode violently, and if so, how? Are there detectable signatures of life elsewhere in the universe? What emerged from the Big Bang, and how did galaxies, stars, planets, and life form and evolve from it?

In the 1970s, the evolution of detectors away from photographic plates and toward highly sensitive electronic chips improved faintness detection limits and the ability to easily archive data, so much so that observing and data reduction techniques of modern times bear little resemblance to those practiced only a generation ago. In the 1980s, impressive advances in optics from laser alignment techniques enabled smaller mirrors to be mosaicked together into a much larger assembly, acting as a single mirror and greatly increasing the light-gathering power of a telescope. Further breakthroughs in the 1990s saw the development of deformable mirrors in telescopes whose shapes could be changed to compensate for the dispersion of light by the Earth's atmosphere. Using the coherent light from a laser, telescopes could now compensate for the rapid fluctuations in the atmosphere that cause blurring of images. On timescales of less than 1/10th of a second, the shape of a flexible mirror in the telescope can be changed to correct for atmospheric distortions; it is effectively as if the atmosphere were not there.

Although it now shares its domain in astronomy with supercomputers and gravitational wave detectors, the telescope remains the primary research facility for astronomers, as it has since Galileo first trained his upon the sky. However, the most essential component of telescopes, the main primary mirror, has evolved in several ways since Newton first constructed a telescope using a mirror. Telescopes are much larger now because of new technology that has created ways by which the primary "mirror" can take on sizes that approach 40 m in diameter via the tiling together of smaller mirrors that function in concert with each other as a single integrated unit. The Keck telescopes on Mauna Kea were the first large telescopes built in this manner; the main mirror of one is shown in Figure 1.4. They each have a primary mirror comprised of 36 individual hexagonal mirrors, each 2 m in size. The individual hex mirrors are aligned extremely accurately with each other via a laser alignment system, so they perform as if they were a single solid piece of glass. This same concept was initiated earlier for radio telescopes, though the alignment process is different. The fact that atmospheric distortion is far less for radio waves than visible light allows the separate reflectors for radio telescopes not to be contiguous with each other, as they are for the Keck telescope, but rather separated by many hundreds of meters, as they are for the ALMA radio array on the Chilean *altiplano* shown in Figure 1.5, or even thousands of kilometers apart. The concept of distributing mirrors—or reflector "dishes," in the case of radio astronomy—far apart has been essential to our ability to resolve finer detail in the objects we observe.

Telescopes that focus X-rays have now been fabricated, requiring complex new technology because X-rays can only be reflected by a smooth metallic or glass ceramic surface when they strike the reflective surface at a very high angle, e.g., near grazing incidence (>80°). This has led to a novel design for X-ray telescopes where the focusing mechanism consists of concentric nested cylindrical parabolic and hyperbolic conic section segments that focus the radiation to a single point. Figure 1.6 illustrates the mirror assembly of the *Chandra* X-ray telescope, which has often worked collaboratively with the *Hubble* telescope. The concentric mirrors form the core of the X-ray satellite, which has made significant discoveries of celestial objects in the past two decades. Such complex optical systems could not have been fabricated more than a generation ago.

As important as the new modes of telescope mirrors have been to modern astronomical research, an equally great impact on astronomical research has been made by ongoing innovation in the detectors that sense the radiation that the mirrors bring to a focus. If the primary detector in astronomy were still the photographic plate, with its low detection efficiency and non-linear response to

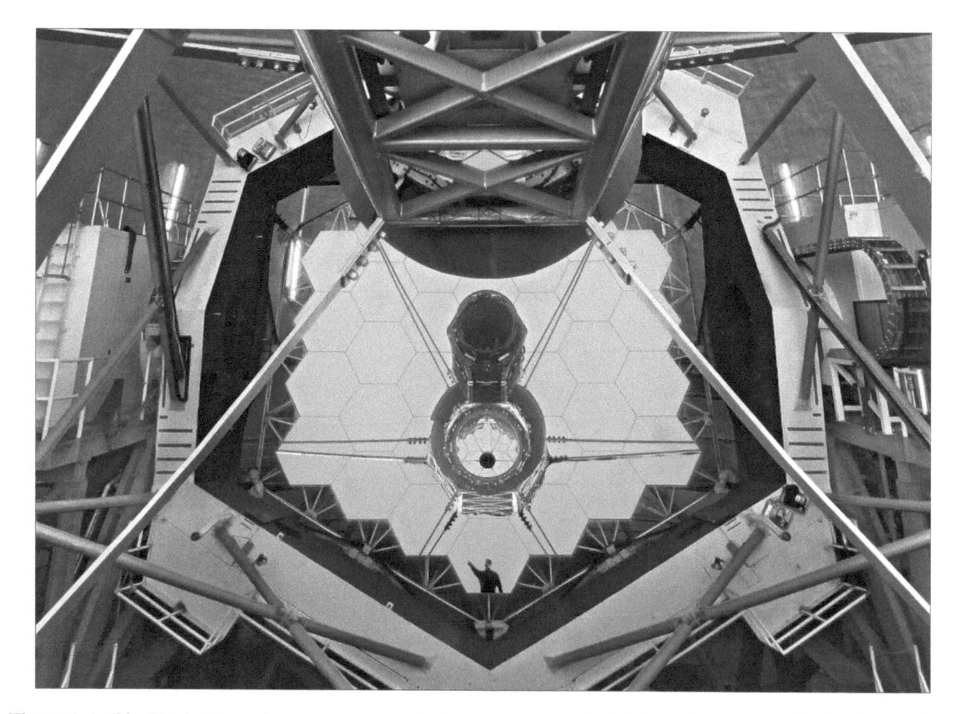

Figure 1.4. The Keck 10 m telescope segmented mirror was easier to fabricate than a monolithic mirror of the same size. The 36 hexagonal segments are aligned precisely using starlight rather than lasers, so they function as if the mirrors were a single piece of glass [credit: Laurie Hatch/Keck Observatory].

Figure 1.5. The ALMA radio telescope array in northern Chile on the Chajnantor *altiplano* at 5000 m elevation. The individual reflecting radio dishes work in concert with each other to constitute one giant radio telescope [credit: ALMA/ASIAA].

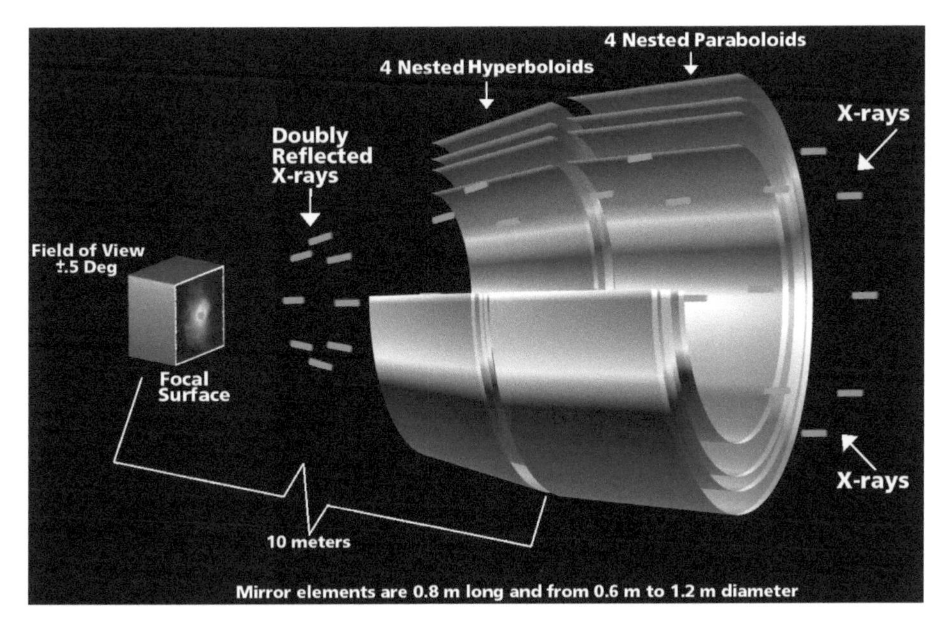

Figure 1.6. A schematic of the nested concentric mirrors for the orbiting *Chandra* X-ray telescope that bring X-rays to a focus on the detector. An unusual geometry for the mirrors is required because solid surfaces only reflect X-rays that strike them at a glancing angle [credit: NASA/CXC].

light intensity, we would still be in the proverbial Dark Ages compared to what we have been able to achieve with data obtained by digital detectors such as the charge coupled device (CCD). The CCD operates on the basis of the photoelectric effect, first explained by Einstein in 1905 and for which he received the Nobel Prize in 1921, by which light produces an electric current when it strikes certain substances. The virtue of CCD detectors is that electric current can be measured easily and accurately, and the current is directly related to the light intensity. The latter is a fundamental parameter that astronomers need to know to determine physical processes in astronomical objects. The photographic plate does not possess any property that enables light intensity to be determined easily, so the introduction of the CCD created a revolution in astrophysics. Without question, it was one of the most important technical advances impacting astronomy in the past century.

Driven by commercial and military interest in imaging devices, i.e., cameras, CCDs were developed in the 1960s by different companies whose research revealed that transistors could be built to capture light much more sensitively than photographic plates or photocathodes. Because transistors can be fabricated quite small, they can be clustered in arrays where each transistor serves as a light sensor and produces its own electric charge and current, thereby serving as a two-dimensional imaging device where each transistor in a pack represents a separate picture element, or pixel. An example of such a CCD that is used on a telescope for astronomical imaging is shown in Figure 1.7. An ingenious way of collecting all the charges, devised by Willard Boyle and George Smith of Bell Labs, provided an efficient electronic process to read the charges. They received the Nobel Prize in Physics in 2009 for their work, and a large industry rapidly sprung up to produce CCDs. These detectors are now universally found in visible-light television cameras and imaging systems worldwide—and certainly in astronomical instruments on virtually all optical telescopes.

Figure 1.7. This CCD, now on display at the NASA Air & Space Museum in Washington DC, was developed for the STIS instrument on the *Hubble* telescope. It consists of 1024 × 1024 individual pixels that detect light far more efficiently than a photographic plate [credit: NASM/Smithsonian].

The ability of CCD pixels to create and store a weak electronic charge that can be read as the proxy for even the faintest light levels detected was essential for detecting faint objects such as those we planned to observe in the HDF. However, as important as sensitive CCD detectors were for the HDF, another crucial factor had to be treated before any deep image could be considered worthwhile—the ever-present faint diffuse background of light against which astronomers observe all objects when looking out into space. When observing with ground-based telescopes, the major source of radiation background is the faint airglow of the Earth's atmosphere, even on the darkest night. During the day, the Sun's radiation energizes the atmosphere, which in turn radiates this energy away. However, it takes many hours for the atmosphere to relax back to a quiescent state. Thus, during the night, it continues to radiate a faint amount of light, including the well-known aurorae, and this radiation will dominate and mask the light from the faintest objects that astronomers are trying to image.

Essential for life and therefore valued by all, the Earth's atmosphere is nonetheless a stumbling block to astronomers—and not just as a radiator of unwanted light that fogs our detectors. The atmosphere not only radiates; it absorbs. Radiation from many wavelengths, including the X-ray, ultraviolet, and infrared regions, cannot penetrate the air we breathe. These regions contain many important diagnostics that astronomers find crucial to their analysis of celestial events. Fortunately, there is a straightforward way to solve the problem of the atmosphere's nuisance role in astronomy—place telescopes in space above the atmosphere.

Easy to contemplate but difficult and expensive to do, launching telescopes into space was a sufficiently important idea that astronomers took great interest in pursuing it when rocketry was developing. The long road followed by scientists that resulted in pioneering space probes and orbiting telescopes like the *Hubble* would not have been feasible were it not for the fact that modern warfare presented society with alternative reasons for taking an interest in rockets fired into space. Along with other technological advances that have advanced science, astronomers benefitted from the history of research and breakthroughs made possible by federal funding of both civilian and defense projects that resulted in the capability to launch large, sophisticated equipment into Earth orbit.

Robert Williams

Chapter 2

Telescopes in Space

As long as telescope mirrors were no larger than a few meters in size and the human eye and photographic plates were astronomers' primary light detectors, the absorbing and blurring effects of the Earth's atmosphere did not present a serious problem for astronomers. However, with the advent of larger telescopes in the latter half of the 20th century and the development of electronic detectors that could sample X-ray, ultraviolet, and infrared wavelengths, all of which are absorbed by the atmosphere, astronomers realized they must make serious efforts to launch telescopes above the Earth's atmosphere. The groundwork for using rockets for scientific purposes had been anticipated earlier in the century by visionaries such as Robert Goddard, a scientist and engineer who chose to work in measured isolation with his small team, largely without public funding for much of his career. Goddard pioneered the development of rockets and their use for scientific purposes before the First World War.

Goddard's career was unusual as a counterpoint to the modern culture of advancement via publication of one's results. In Goddard's case, it was patents based on his lab work and experiments, even more than the publication of his 1919 monograph *A Method of Reaching Extreme Altitudes*, that are his most important legacy. A civilian professor at Clark University in Worcester, MA for part of his career, before re-locating to Roswell, NM due to its more favorable launch conditions, Goddard worked outside of military efforts devoted to rocketry and he was unswerving in his general disinterest in exchanging ideas or communicating his experiences with other researchers. He was convinced he could achieve more rapid turnaround time on projects and also protect himself from "distracting" criticisms that he expected to (and in fact, did) receive from the media, by operating a largely closed shop. His important successes, such as his pioneering development of liquid-fueled rockets at a time when the rest of the world was focused on solid-fuel engines came after decades of hard work. They eventually caused him to evolve to a more collaborative style that was instrumental in his receiving research funding from

doi:10.1088/978-0-7503-1756-6ch2

Figure 2.1. Robert Goddard developed the world's first liquid-fueled rockets. This prototype was launched in 1926 in Massachusetts, travelling 60 m in 2.5 s [credit: NASA].

the Guggenheim Foundation in the latter stages of his career, which provided the funds needed for his experimental launches that were important for the development of U.S. rocketry (Figure 2.1).

Following World War II, rockets were developed for civilian purposes that involved scientific experiments and remote sensing, including opportunities for launching small payloads that contained telescopes. In the late 1940s and early 1950s, the launching of orbiting satellites was not yet possible, so space efforts were devoted primarily to rocket flights that reached maximum altitudes of hundreds of kilometers above the Earth's surface and lasted less than an hour. Stabilizing the rocket so a telescope could be pointed to a fixed position of the sky was a real challenge with the modest computers and technology of the time, so meaningful astronomical observations from a rocket were very limited. An increasing civilian space industry was developing, however, and with it were associated planning and policy exercises carried out at *think tanks*. The Rand Corporation was one such organization.

Named from the contraction of "***r***esearch ***an***d ***d***evelopment" the Rand Corporation was a subdivision of Douglas Aircraft Company at the end of the war, dedicated to connecting military planning with research and development decisions. In 1946, it was selected by the Air Force to undertake a study of the utility of artificial Earth satellites, which were considered to be achievable in the foreseeable future. As a prelude to the study, scientific advisors to U.S. Secretary of War Henry Stimson contacted astronomer Dr. Lyman Spitzer, who was known to them from his directorship of the wartime Sonar Analysis Group, to suggest possible uses of orbiting satellites for astronomy. At the end of the war, Spitzer was in the process of returning to his professorship at Yale University, where he had been on leave of absence while seconded to the wartime Special Studies Group, and he was interested in putting astronomy forward as a legitimate priority on the national space agenda. Spitzer agreed to spend one week at Douglas Aircraft in Santa Monica, and contribute to the satellite study. During that time, he wrote a four page addendum to the Rand Corp study entitled *Astronomical Advantages of an Extra-Terrestrial Observatory* that ultimately would have a profound influence on astronomical activity in space (Figure 2.2).

Spitzer's advocacy of space astronomy was built upon a concept proposed in 1923 by Hermann Oberth of Germany, who envisioned using a rocket to propel a telescope into space. Oberth first advocated the idea in a paper entitled *The Rocket in Planetary Space* that was based upon work he had done for his research dissertation at the University of Heidelberg—and which was rejected because it was deemed "impractical." Although Oberth's primary interest was in rocketry rather than the scientific value of a telescope in space, the fact is the idea of a space telescope was born in his writings. Later in his career, Oberth took on a young Wernher von Braun as his assistant and they worked together on the German V2 military rocket. Following the war, they collaborated on the development of large rockets for military and civilian purposes in the USA. By contrast, Spitzer's essential contribution to the concept of a telescope in space was to highlight the important *scientific* advances in astrophysics that would come from orbiting a large telescope that could be devoted to addressing such crucial topics as the structure of galaxies and the nature of other planets in the solar system. Although obvious to us now, the idea of a large space telescope was so far ahead of its time in the mid-1940s that Spitzer's report was hardly noticed by the scientific community while more achievable goals were pursued.

At the time of the post-war Rand Corp report, the German V2 rocket was the world's most advanced, yet its capabilities were quite limited. It often failed at launch, and even when it succeeded, its guidance system was rudimentary. Still, it was clear that rocketry would be essential in the future, both as a weapons delivery system and for its ability to put satellites into orbit that could benefit society, e.g., telecommunications satellites, as well as for its use for launching probes to explore the solar system. These forces motivated the development of increased launch capabilities over the next two decades that consisted of the launching, sometimes unsuccessfully, of a mix of rocket flights that had different purposes. The smaller launches were above-atmosphere civilian missions that carried scientific equipment and were notable for their success in achieving significant astronomical breakthroughs.

The Astronomy Quarterly, Vol. 7, pp. 131-142, 1990
Printed in the USA.

0364-9229/90 $3.00+.00

ASTRONOMICAL ADVANTAGES

OF AN

EXTRA-TERRESTRIAL OBSERVATORY

LYMAN SPITZER, Jr. [1]

This study points out, in a very preliminary way, the results that might be expected from astronomical measurements made with a satellite vehicle. The discussion is divided into three parts, corresponding to three different assumptions concerning the amount of instrumentation provided. In the first section it is assumed that no telescope is provided; in the second a 10-inch reflector is assumed; in the third section some of the results obtainable with a large reflecting telescope, many feet in diameter, and revolving about the earth above the terrestrial atmosphere, are briefly sketched.

It should be emphasized that this is only a preliminary survey of the scientific advantages that astronomy might gain from such a development. The many practical problems, which of course require a detailed solution before such a satellite might become possible, are not considered, although some partial mention is made of certain problems of purely astronomical instrumentation. The discussion of the astronomical results is not intended to be complete, and covers only certain salient features. While a more exhaustive analysis would alter some of the details of the present study, it would probably not change the chief conclusion -- that such a scientific tool, if practically feasible, could revolutionize

---

[1] The report re-printed here appeared as Appendix V of a larger document prepared for the Project RAND of the Douglas Aircraft Co., on 1 September 1946. At that time, Prof. Spitzer was on the astronomy faculty of Yale University; he has been affiliated with the Princeton University Observatory since 1947.

Figure 2.2. The title page of Lyman Spitzer's 1946 addendum report for the Rand Corporation on the astronomical uses of a space telescope (Reprinted from Spitzer, L., Jr., Report to project rand: Astronomical advantages of an extraterrestrial observatory, AstQ, 7.3, 131, Copyright 1990, with permission from Elsevier.)

Figure 2.3. Launch of a modified German V2 rocket from White Sands Proving Ground in 1949. Numerous scientific observations and experiments were conducted on sub-orbital flights from White Sands, beginning in the late 1940s [credit: White Sands Museum/NASA].

In the decade following Spitzer's 1946 report, a progression of astronomical observations were carried out at X-ray and ultraviolet wavelengths from rockets that successfully detected radiation from the Sun, objects in the Milky Way, and a collection of very energetic galaxies. The first detection of cosmic X-rays (from the Sun) occurred in August 1948 from an instrument launched aboard a V2 from White Sands Proving Ground (Figure 2.3). Larger, more complex launches were important technology demonstrations whose payloads consisted of experiments and telecommunications satellites. What resulted was a stream of technological advances that made the launch of larger payloads, eventually including astronauts, feasible. These were undertaken beneath the umbrella of the National Aeronautics & Space Administration (NASA), a large federal agency that advocated and coordinated non-military space activity for the U.S.

NASA was created by an act of Congress in 1958, out of a sense of urgency spurred by the Soviet Union's successful orbiting of Sputnik the previous year. It had become clear even before the Cold War began that space would inevitably be the ideal theater for the delivery of nuclear weapons and the U.S. needed to establish a prominent presence in space for military reasons. In addition, as advocated at the time by President Eisenhower, the scientific and technological advances that would come from maintaining active programs in space made a U.S. civilian space program compelling. The primary mission of NASA, mandated by its Congressional charter, was to foster U.S. leadership in peaceful activities above the Earth's atmosphere, and astronomy was designated as one of the included disciplines.

Figure 2.4. NASA's Dr. Nancy Roman with an early model of the Large Space Telescope that was eventually developed as the *Hubble Space Telescope* [credit: Emilio Segre Visual Archives/AIP/Science Photo Library].

The initial staffing of NASA was accomplished, in some haste, by the transfer to the new agency of volunteers who possessed the necessary skills and expertise from associated governmental agencies like the Naval Research Lab (NRL). One such NRL transfer was astronomer Dr. Nancy Roman, who qualified as an experienced researcher because of her experience with NRL rocket flights. After a series of interviews with NASA administrators, she was offered the lead position of the new astronomy section at NASA headquarters, with a mandate to create a program for astronomy. Dr. Roman worked closely with external advisory committees, whose members came from academia and industry, toward the formulation of the initial NASA astronomy program (Figure 2.4). What they crafted was a program that proposed for NASA to launch a series of telescopes having primary mirrors with diameters on the order of 1 m, within the next 3–6 years. Eventually named the Orbiting Astronomical Observatories (OAOs), NASA did follow through in 1960–61 by issuing requests for proposals to build the OAOs. Among the successful groups that were eventually selected by a peer-review process to lead OAO missions were the University of Wisconsin and Princeton University, along with collaborating institutions.

Four OAOs were constructed and launched in the period 1966–72, and they constituted the major part of the NASA astronomy program for that period. Goddard Space Flight Center (SFC) outside of Washington, DC, was assigned the major role of managing their development. All of the OAOs operated instruments that explored the ultraviolet region of the spectrum, which cannot be accessed from the ground. Of the four satellites, two suffered major failures associated with their launches, whereas the other two operated well and produced excellent science.

Even with its successes during this period, the NASA astronomy program was not NASA's paramount mission, when one considers the agency's achievements in advancing supersonic flight, implementing the system of telecommunication satellites that revolutionized worldwide communication, and placing humans on the Moon. Nonetheless, the launches of the Ranger and Mariner planetary probes to study the Moon and Mars stirred great interest in the public and were significant scientific and public outreach successes for astronomy. However, it was the decision by NASA in 1972 to move forward with the Space Shuttle that gave a jump start to a much larger, more complex astronomical telescope in space.

The Shuttle program came about as a result of a request made by President Nixon to NASA in 1969, that it formulate a strategic plan for a U.S. presence in space after the Apollo lunar missions. A task group headed by Vice President Agnew was created to assess plausible future directions for NASA. Their final report to the President, after a year's work, was ambitious. They recommended: (1) a low-Earth-orbit space station, (2) a manned mission to Mars, and (3) the development of a reusable space shuttle transportation system (STS) that would enable payloads and astronauts to be launched into Earth orbit every few weeks. Because of the large costs associated with each of the recommended programs, Nixon accepted only the shuttle program as being realistic; he proposed it to Congress for approval in 1972 January. Itself expensive in terms of NASA's historical budgets, the Shuttle program was hotly debated by Congress, but approved by the end of that year. It promised to be a game changer in providing the scientific community with frequent launches and the possibility of long-term maintenance of experiments in a space environment.

The two successful OAO missions that produced solid scientific results had a dramatic effect on many astronomers, leading them to realize that the early vision of Lyman Spitzer for a large space telescope was not only attainable but could yield discoveries that were not possible from ground-based telescopes. Lobbied by astronomers and the contractors who had built components for the OAOs, NASA authorized several studies that demonstrated that a telescope of aperture 2–3 m was technically feasible and could be a cornerstone of NASA science in space. The idea gathered momentum within NASA as the Space Shuttle program was being put forward, and it took on increased importance as part of the Shuttle program because it was a natural fit for the Shuttle's capabilities.

The persistent efforts of astronomers who had developed a solid working relationship with NASA from the early rocket flights and OAO missions, as well as the credibility of astrophysics within the National Academy of Sciences, the primary organization advising Congress on matters of science, were key factors in achieving what eventually turned out to be *Hubble Space Telescope*. The process by which large, expensive astronomical projects are approved and developed is complex. Projects that involve the major federal funding agencies, i.e., NASA, the NSF (National Science Foundation), and the DOE (Dept. of Energy), are subjected to a two-pronged process that requires extensive scientific community vetting and approval via a process initiated at the beginning of every decade by the National Academy. It also requires approval by the federal funding agencies, whose priorities are set in the different fields by their own committee processes. In short, the procedure by which large projects are approved is lengthy and complicated.

Once the federal agencies accept projects as part of their program plans, they negotiate with Congress over the funding of the projects—an intense interaction unto itself. This system can be unwieldy and bureaucratic, and it is certainly influenced by disparate personalities that have distinct visions. The process has evolved in a sensible way over the years to merge the desires of scientists with the political realities mandated by Congressional funding. When one looks at the important discoveries that have come out of the process, the successes that have resulted in significant advances predominate over those projects that, for whatever reason, were never approved to move forward.

The idea of a large telescope in space was subjected to this process beginning in the 1970s. The first National Academy of Science decadal report for astronomy, issued in 1972 and chaired by Caltech astronomer Jesse Greenstein, did consider and recommend a large space telescope, but only as a second-priority project and for a future decade because of budgetary concerns. The idea of the Large Space Telescope (LST) had emerged during the OAO era and was steadily refined over time as conditions and technology changed, eventually emerging as the multi-purpose 2.4 m diameter *Hubble Space Telescope*.

During the time the OAOs were the focus of US space astronomy in the 1960s, the idea of larger space telescopes received serious attention from several groups: NASA's Space Studies Board, the National Academy, private contractors who supplied components for the OAOs, and the consortium of universities that form the Association of Universities for Research in Astronomy (AURA) all conducted feasibility studies in committees and working groups for space telescopes substantially larger than the OAOs. For different reasons, these studies gained little traction within the astronomy community. In addition to technical issues that needed to be solved, a number of astronomers in leadership positions called attention to the great cost that pursuing a large space telescope would require, possibly disadvantaging other areas in the field for the future. On the other hand, a significant benefit of the LST studies was the fact that they did stimulate studies and the development of prototypes that resolved some of the issues that would be faced by an LST, e.g., its ability to retain pointing stability while moving in orbit.

NASA encouraged and partially funded studies as a means of maintaining the possibility that a large telescope might be an important project for the future. Interest in a future LST was particularly keen at NASA Goddard SFC, based on its prior experience with the OAOs, and at Marshall SFC, driven by the presence there of its first director, Wernher von Braun, who saw the LST as a major program that fit Marshall's expertise. One study commissioned by NASA to Grumman Industries, who were the prime contractor for major components of the OAOs, tasked them with determining the costs of orbiting an LST with an existing launcher as opposed to using the Space Shuttle. Grumman's 1970 study concluded that the

Shuttle would be significantly cheaper to use as the orbiter, and doing so would enable periodic maintenance of the telescope and its many sensitive electronic components.

A number of elements came together in the early 1970s that had significant impact in making the LST concept a reality. The history of what finally morphed into the *Hubble Space Telescope* involved a complex web of events that have been documented in several reference books[1]. We confine our attention here to those elements of the *HST* that were key for the conduct of science with the telescope. Of particular importance were several major decisions by NASA at the beginning of the process, the most important of which was the decision to fund independent studies of the feasibility of an LST and to assign responsibilities to several NASA centers to have control over the major components of LST development. These included its operation, data handling, and instrumentation, as well as the interface between NASA, its contractors, and the science community. NASA headquarters assigned Marshall SFC responsibility for the basic telescope assembly while Goddard SFC was assigned responsibility for the instrumentation, post-launch operation of the facility, and primary interface with the science community.

At NASA headquarters, Nancy Roman had been the driving force for an LST from its earliest days. As momentum built for the project (which was not officially approved yet by NASA), it arrived at a point where a dedicated Project Scientist was needed to garner support and sell the project to the astronomical community. In 1972, University of Chicago/Yerkes Observatory Director C. R. O'Dell was recruited to take on this task, with one of its principle challenges being that it required interfacing with three very different cultures: NASA, private industry, and the academic community. For more than a decade, Roman worked with O'Dell, whose position was officially assigned to Marshall SFC. They were the primary advocates and overseers of the LST within NASA, and their efforts working with the astronomical community produced a detailed paradigm for NASA operation of a large scientific project that now serves as a standard for large astronomical facilities.

Several important features that have been a crucial part of the scientific success of *Hubble Space Telescope* came about from NASA's partnership with the astronomical community during the development of the LST. The manner in which teams based in academia were able to develop the instruments, in an era when detector technology was changing rapidly, produced excellent instruments. In addition, the creation of an independent institute, external to NASA, that was given responsibility for managing the scientific program of the telescope was a new approach that broadened the way science was conducted on NASA missions. Without either of these essential parts of *HST*, it is questionable whether the Hubble Deep Field or any of its successor deep field surveys would ever have been attempted.

The involvement of academic institutions in building instruments for telescopes in space began in the 1960s with the OAOs, where universities worked closely with

[1] *The Space Telescope* by Robert Smith (1989; Cambridge: Cambridge Univ. Press), and *The Universe in a Mirror* by Robert Zimmerman (2008; Princeton, NJ: Princeton Univ. Press).

primary private contractors, such as Grumman Aircraft, to develop major components for the satellites. The complexity of a large space telescope made its instrumentation a huge undertaking. Not the least important was the fact that technology was changing so rapidly that revisions to optimize the instrumentation were constantly occurring, thereby motivating changes to all telescope components that interfaced with the instruments. Given the long lead times required for flight readiness and necessary reviews, this led to constant struggles within every part of the project. NASA wanted a coherent, viable project that would advance in an orderly fashion, as did the astronomers. However, the astronomers were less concerned about bureaucratic details and deadlines than they were about having state-of the-art instruments and detectors on what they were hoping would be the discovery telescope of their lifetimes.

A key element affecting LST instrumentation was the evolution of astronomical detectors away from photographic plates and toward digital detectors. The latter, in the form of CCDs, were much more sensitive to light than plates, and they have the advantage of having a linear, easily measurable response to radiation that makes quantitative measurement far easier than the complex processes that were required to convert a photographic image to a brightness level. The earliest studies of a large space telescope in the late 1960s proposed astronauts changing out bulky caches of photographic plates in orbit. Fortunately, within two decades, those plates would be replaced by small solid-state electronic CCD detectors whose measurements could be stored electronically and transmitted directly to the ground.

The transition from photographic plates to CCDs involved a stage where image intensification systems, viz., SIT and SEC Vidicon systems, were used in both ground and space facilities. These represented large improvement over plates, especially in their ability to detect ultraviolet light. Such imaging systems were used on the OAOs and Apollo lunar missions, but they had their own limitations. As a result, various forms of detectors were advocated by different instrumentation groups that were providing space telescope instruments, and this required a great deal of attention and decision-making by NASA and advisory committees.

The management role of the LST project scientist Bob O'Dell—who, as both a research scientist and NASA manager, understood the importance of both sides of this issue—was crucial in keeping energetic forces channeled productively. Able to be an advocate for either side at different times, he served as a buffer in the collaborative versus competitive relationship between NASA and the researcher instrument builders. To the credit of NASA, they realized how important advances in detector characteristics were to the eventual success of orbiting telescopes, so they also understood the importance of giving the instrument teams leeway despite severe budgetary constraints. In fact, the involvement of the academic community in instrumentation development was crucial to getting astronomers on board as vocal advocates for the LST, before it had official approval by NASA and funding from Congress. The suite of instruments that have been installed on the *Hubble* do represent the state of the art, as much as the lengthy process of flight readiness and the difficult space environment allow, and they have performed excellently. Their digital nature, converting light photons into electric charge, has made the detection and accurate measurement of faint astronomical sources possible, and they have been the scientific backbone of *HST*.

As a forefront scientific facility, the *Hubble* telescope requires technically advanced instruments that can detect the light from cosmic objects in different ways that reveal their true nature. The instrumentation of any telescope is crucial to its performance; without instruments. a telescope remains a simple light bucket. The maintenance, updating, and replacement of the instruments are paramount to the telescope's ability to make discoveries. The location of the *Hubble* in a low Earth orbit that allowed the Space Shuttle to reach it for periodic servicing was crucial to the success of the telescope. Realizing this from the beginning of the project, core groups of astronomers were always present at planning meetings; they were motivated to develop the most advanced instruments and detectors for the telescope.

An integral part of the development of a large orbiting telescope in space was NASA's decision to hold an open competition for the telescope's instruments. Significant expertise and interest existed within the community when NASA put out an announcement of opportunity and request for proposals for the primary telescope instruments in 1977 March. This was only months after NASA had designated the Marshall and Goddard SFCs to provide management and oversight of an LST project. A number of proposals were submitted by groups with experience in building instruments for ground-based telescopes, and these were evaluated by an independent board in mid-1977. Noel Hinners, head of the Office of Space Science, received the recommendations of the board late in the year, and selected the instruments and their development teams. The winning proposals consisted of: two imaging cameras, the Wide Field/Planetary Camera (WFPC) as shown in Figure 2.5

Figure 2.5. The original /WF/PC, developed at the Jet Propulsion Lab, that obtained most of the early images from the spherically aberrated *HST* before being replaced by WFPC2 on the first *Hubble* servicing mission [credit: NASA/JPL].

and the Faint Object Camera (FOC); two spectrographs, the Faint Object Spectrograph (FOS) and the Goddard High Resolution Spectrograph; and the High Speed Photometer (HSP). Science on the LST, as it was called at that time, would be carried out with these instruments.

One of the unique features of the *HST* is the existence of a separate, independent institute that conducts the science program for the telescope under contract with the HST Project at Goddard SFC. The science programs of previous NASA space science facilities had normally been managed in-house by NASA centers. This paradigm was not followed for the proposed LST, for a variety of reasons; perhaps the most important reason was that the astronomy community wanted as much control over *HST* science as possible. To the many astronomers in universities and observatories outside of NASA, optimal science would result from creating an organization independent of NASA over which they could exercise substantial control.

In an important way, the idea of an independent institute had its beginnings in serious budget problems that emerged in 1976, due to the high cost of the Vietnam War, and consequent changes in priorities within NASA for its different programs. Early in the year, when the government released the President's proposed budget for 1977, NASA cut the Space Telescope project entirely out of its budget. NASA Administrator James Fletcher made the decision to direct LST funding to the Space Shuttle program because of what he termed "Shuttle funding requirements" resulting from the high priority Congress was placing on the Shuttle.

In prior years, funding had been provided in increasing amounts for studies devoted to different aspects of the space telescope, and these had given momentum to the LST project. Statements from various NASA managers in support of the telescope had astronomers counting on an official NASA "new start" for LST in 1977, so the sudden plans to cancel LST stunned the astronomical community. One leading astrophysicist, Princeton's John Bahcall, reacted to this unexpected turn of events by organizing an intense lobbying effort that was directed at members of Congress. Astronomers throughout the U.S. were mobilized in a campaign to meet with their members of Congress in order to inform them of the educational and scientific benefits of studying the cosmos from space. The goal was to influence NASA to change its mind about LST and resurrect it. Around this time, the astronomical community realized that it would be wise to drop the word *Large* from the Large Space Telescope project because it might preclude arguing in the future for a larger space telescope. Henceforth, the LST project would be referred to as simply the Space Telescope (ST), to be launched aboard the Space Shuttle.

The campaign to resurrect the Space Telescope was an unqualified success. After months of coordinated effort, the ST was restored as a NASA priority, and the campaign had the added benefit of having educated large numbers of astronomers about the telescope, turning them into strong advocates for it. The successful campaign did produce some significant changes in the NASA plans for the ST, however. In previous years, both Congress and the Office of Management and Budget (OMB) had insisted that NASA make the ST project international by finding foreign partners that would share costs. NASA had seriously approached the European Space Agency (ESA) about ST participation and the two agencies were close to an agreement that would make ESA an official partner in the program at a 15% level. As a condition for having funds reinstated back into NASA's 1977 budget, NASA and ESA did sign an agreement that called for ESA to provide an instrument and the solar arrays for the telescope, and to contribute scientists and technical expertise to the ST effort. The official involvement of European scientists in all aspects of the telescope project, including science policy committees, had a major impact in making the space telescope well-known outside of the U.S.

The year-long battle to restore ST within NASA's budget had a telling influence on the astronomical community. Astronomers woke up to the fact that, when they united in action, they were able to have a significant impact on the development of a very large national project. When NASA subsequently restored the ST to its budget for FY1977 in the form of an official new start, they suddenly found an energized and educated community that now wanted a say in major ST matters—particularly its instruments and the way in which its science programs would be determined. To a large extent, the astronomy community had been influenced by the success of a paradigm that had been established several decades previously for optical ground-based astronomy, where universities had formed consortia to operate national centers that enabled the development of large telescopes.

Kitt Peak National Observatory had been formed by the AURA, Inc., university consortium: to many astronomers, this presented a good model for the ST to follow, as opposed to having NASA centers over which they had much less leverage to manage conduct of the science. The concept of having national centers for astronomy was appealing to a large segment of the astronomical community that was associated with smaller colleges and universities who did not have access to forefront facilities. A national center whose facilities were available to all, regardless of university affiliation, was a way any scientist could undertake important science even if their particular college/university could not provide for the most modern equipment.

With the support of the astronomy community, John Bahcall and Bob O'Dell lobbied NASA hard for the creation of a separate institute, external to NASA, to administer the full science program for the ST. Not surprisingly, the Goddard and Marshall NASA centers associated with the ST project opposed the idea. Not only did the concept of an independent external institute go against the NASA tradition of their centers managing science programs for their missions, but the centers were doing a very good job administering the science of other major missions, such as the International Ultraviolet Explorer. Significantly, however, within the top levels of NASA, there were several key players with ears sympathetic to the idea of an independent institute.

Both Nancy Roman and Noel Hinners, head of the Office of Space Science, reacted favorably to the community desire for an independent institute. Given NASA's long-term commitment to the *development* of major programs like human space flight, Hinners was particularly worried about continuing NASA interest in ST *operations* after launch, when budget pressures could easily cause them to lose interest in the ST. An independent institute attached to an external organization could be more effective in keeping pressure on NASA for the continuation of operations for a productive science program. Thus, in 1976, Hinners had the insightful idea of asking the National Academy of Science (NAS) to conduct a study of the pros and cons of creating an independent science institute for the ST and to make their recommendation to NASA, realizing that Congress and the Office of Management and Budget (OMB) would listen to the NAS.

The NAS did constitute a committee, chaired by President Johnson's Science Advisor Donald Hornig, to assess the value of creating a separate science institute that would operate under contract to NASA but act independently of it. The committee submitted its report to the NAS, affirming the wisdom of having the ST science program conducted by the external institute. NASA ultimately accepted this recommendation and initiated a process to create the institute, resulting in a request for proposals for the institute in 1979. The university consortium AURA, Inc., submitted a proposal to create an institute at Johns Hopkins University (JHU) in Baltimore, and it emerged victorious in the competition. Following negotiations, NASA authorized AURA to establish the Space Telescope Science Institute (STScI) in 1981.

AURA began work immediately to create STScI, negotiating with JHU to provide space for the Institute. It defined its responsibilities with Goddard and appointed a founding director, Riccardo Giacconi, an experienced X-ray astronomer who had been a strong advocate for national centers in different areas of astronomy. Early in Giacconi's tenure at the Institute, NASA decided it was appropriate to give a more specific name to the Space Telescope. They appointed a largely in-house committee that conducted its business of considering possible names for the Space Telescope out of the public eye. In 1983, with little fanfare, the committee submitted its recommendation to NASA's top administrators via an internal memo: re-name the ST the *Hubble Space Telescope,* after American astronomer Edwin Hubble, who with Belgian prelate-astronomer Georges Lemaitre, first established the nature of the universal cosmic expansion.

The creation of the STScI in Baltimore (Figure 2.6) has had an important impact on the way science has been conducted with the telescope during the more than two decades it has operated in orbit. This is not to say that the discoveries from the telescope would have been different had the independent Institute not been founded and the *Hubble*'s science program were instead managed by a group within NASA at one of its centers. The *Hubble* telescope's capabilities have been so unique that forefront science was guaranteed to come out of its observations. However, the manner in which the science has been carried out by an external organization that is directly overseen by the international community has enabled it to implement procedures free of constraints imposed by the framework of a government agency and tighter federal regulations.

The STScI has had responsibility for conducting the science program of *HST* and has done so under contract with the *HST* Project at Goddard SFC. Goddard has retained responsibility for the health and safety of the *HST* spacecraft crew.

Figure 2.6. The Space Telescope Science Institute, which conducts the science program on the *Hubble Space Telescope*, on the campus of Johns Hopkins University in Baltimore [credit: STScI].

The two organizations, STScI and Goddard, have maintained a close relationship that has been remarkably productive and collaborative, even through tense times resulting from budget problems and equipment malfunctions on the telescope.

AAS | IOP Astronomy

Hubble Deep Field and the Distant Universe

Robert Williams

Chapter 3

A Flawed Hubble Telescope

On 1990 April 24, the NASA/ESA *Hubble Space Telescope* was launched from Cape Canaveral onboard Space Shuttle Discovery, along with a seven-person crew. After achieving orbit and performing equipment checks, the deployment of the telescope went smoothly. Compared to the more complicated servicing missions to the telescope that would take place in subsequent years, requiring delicate work within the interior of the telescope structure, the deployment mission was straightforward. On April 29, the *HST* was released from the shuttle while Discovery's thrusters gently backed the shuttle away from it. After more than 20 years of development, the *HST* was on its own and ready to make observations (Figure 3.1).

The launching of the *HST* following its official NASA new start in 1977, its assignment to Goddard as a cornerstone project in NASA's Great Observatories program, and the creation of STScI to manage its science program were all the fruition of decades of planning and effort. It involved a large network of contractors, led by Lockheed Martin Space Systems, who built the main telescope assembly and managed the integration of the many parts of the telescope. The instrument development was overseen mostly by collaborations of university scientists, with companies such as Perkin-Elmer and Ball Brothers doing much of the fabrication and system integration. Operations software was contracted out to different companies, and STScI was responsible for coordination and oversight of the software development. The complex *HST* Project was an enormous challenge for program management: at the time of launch, it carried a total cost of more than $2 billion, making the *HST* the most expensive scientific project in history at that time.

Originally scheduled for launch in 1986, *HST* deployment was delayed four years by the Challenger disaster, which interrupted all Shuttle launches while modifications were made to the Shuttle engines and solid rocket boosters. In some regards, the delayed *HST* launch had fortuitous consequences for the Project, in that it allowed time for the refinement of operations software that had still been in a rudimentary state at the time of the scheduled 1986 launch. The subsequent

doi:10.1088/978-0-7503-1756-6ch3

Figure 3.1. The *Hubble Space Telescope* at the moment of its release from Space Shuttle Discovery in 1990 April. The aperture cover remains closed to prevent contamination from the Shuttle's rocket thrusters as it moves away from the telescope [credit: NASA].

four-year delay, while bringing huge budget pressures to NASA, did serve to put the entire project and telescope assembly on a firmer footing. What NASA and the world were about to find out, however, was that the *HST* had a major, debilitating flaw, which was about to make *Hubble Space Telescope* a household name . . . to be scorned and pilloried.

The *Hubble*'s first month in orbit consisted of a complex schedule of letting the telescope assembly outgas, bringing up the electronics, and aligning the optical train of the telescope and instruments. The solar arrays, batteries, communication antennae, gyroscopes, reaction wheels, computers, tape drives (before the digital age had arrived in space) all needed to be activated and tested. Instrument pickoff mirrors, filter wheel assemblies, focusing mechanisms, and detector electronics all needed to be put through processes that would make them operational. In particular, testing of the secondary mirror alignment and focusing mechanism at the open end of the telescope was crucial to bringing the light into proper focus in the sharpest detail for the instruments.

The quiet, passive outgassing of the telescope was one of the important aspects of the two-month Orbital Verification phase of the *Hubble* following its launch. Made of modern synthetic materials, such as carbon fiber epoxy, the telescope tube structure would outgas in the vacuum of the orbit. The release of gas produces small changes in the length of the tube that cause the secondary mirror, whose connecting struts are anchored into the tube, to move and therefore change the focal point of the telescope. A month of outgassing was sufficient to produce a relaxed, more stable telescope structure, such that frequent refocusing of the *Hubble* would not be necessary.

There was an additional threat to the *Hubble* that outgassing would cause: the mirror could be contaminated by escaping gas if accidental mispointing of the telescope allowed sunlight to strike any part of the interior of the telescope tube assembly. In the same way that automobiles parked in direct sunlight for several days with all their windows closed develop a noticeable white film on the inside of the windshield, the same phenomenon would happen to the *Hubble* mirrors, seriously compromising their ability to reflect light. In cars, the windshield film is produced by sunlight hitting the synthetic material of the dashboard, generating increased outgassing in the form of complex molecules called polymers. Polymers tend to stick to finely polished surfaces such as glass, and then they absorb and scatter light that strikes the glass. Were this to happen to the *HST*, the light throughput, especially in the ultraviolet, would be severely reduced. Telescope tube outgassing during orbital verification was essential to avoiding any future contamination of the telescope and instrument optics.

The most important goal of orbital verification was obtaining the first images, and that was done as part of positioning the secondary mirror for optimal focus. Not surprisingly, no other task created as much interest as receiving the first images from the telescope. A procedure was carried out that involved checking the WF/PC performance with the secondary mirror in its launch position where optimal focus was expected, after which the first exposure would be attempted. A group of NASA, WF/PC instrument team, and STScI personnel were present at both Goddard and STScI as the first image of a loose group of stars was taken and received on their computers on Sunday, May 20; a fateful day for space astronomy. What followed is a compelling story that has been documented in more detail in the Robert Smith and Robert Zimmerman books, cited previously.

There were varying initial reactions to the first image, that of a loose star cluster, from the groups that were located at the two sites to witness the first exposure. That image, shown in Figure 3.2, did show distinct stars with bright cores surrounded by faint halos. At Goddard, where the media were present, there was a feeling of gratification that a credible image had been obtained. There was no expectation that the telescope focus was necessarily at its ideal position and the fuzzy halos were only prominent if the digital image was displayed magnified. Examining the first image at the same time at the Institute and out of the public eye, optical experts studying the image had somewhat different opinions. On one hand, the image demonstrated that the data stream worked satisfactorily and the image was not overtly distorted. On the other hand, experienced observational astronomer Roger Lynds noted that when the digital display of one of the brightest stars in the field was examined closely, as displayed in the figure inset, the halo showed an unusual structure that he sensed might be caused by something other than just the telescope being out of focus. The *HST* Project people down at Goddard did not note specific image features that they considered to be anomalous, so they advised journalists present that *Hubble* telescope had produced a good image and that refinements in telescope focus would be made that would produce sharper images. The *HST* Project believed the optimal

Figure 3.2. A portion of the first image, of the star cluster NGC 3532, taken by *HST* on 1990 May 20 by the WF/PC. The inset, from a subsequent WF/PC image, shows the magnified image of a star with the spherically aberrated halo seen clearly [credit: NASA/ESA/STScI].

telescope focus was not quite where the telescope mirror had been positioned when the image was taken, and this would be adjusted easily.

Operational meetings were planned at Goddard every day during the two-month orbital verification period. At the Monday meeting following the Sunday "first light" image, the optical quality of the image was discussed. When it was noted that the star images did not appear as sharp as had been expected, Roger Lynds commented that the halo structure signaled to him the possibility of an optical aberration. His concerns were dismissed by the NASA people running the meeting because of their belief that the fuzziness of the images was simply due to poor telescope focus from the incorrect position of the secondary mirror. Poor focus could be improved easily by moving the secondary mirror to its proper location via an attached motor, whereas an aberration would have grave implications with no easy solution: it would mean that the *Hubble's* mirrors were misshapen.

Understanding the image quality of the telescope became a high priority among the different groups tasked with obtaining and analyzing *HST* images. These included NASA operations personnel, the WF/PC instrument team, and scientists at STScI. All wanted to understand the detailed structure of the images, i.e., the so-called *point spread function* that represents the detailed image structure of a point source star on the WF/PC instrument CCD detector. The groups approached the problems in different ways, and in the weeks following first light, both NASA headquarters and the *HST* Project at Goddard were understandably committed to ensuring that any results coming from the different groups would be fully vetted within NASA and the *HST* Project. They had to be substantiated in order to avoid incorrect information about the *Hubble,* which might cause problems with either public or Congressional perception of the $2 billion Hubble Project.

Within several weeks it became increasingly clear that the *HST* optical quality investigations had uncovered seriously bad news that reflected poorly on various decisions and actions that were taken during the development of the *Hubble*. Yet, in spite of all the criticism that rained down on NASA and its contractors from this situation, there emerged numerous defenders who leapt to the fore, both inside and outside of NASA, whose efforts were instrumental in finally making *Hubble Space Telescope* an icon of scientific discovery. In the meantime, the problems with *HST's* first image were about to get the undivided attention of the media in the U.S.

Two of the key people whose efforts first served to clarify the situation with the image were Sandra Faber and Chris Burrows. Burrows, an STScI scientist specializing in optical systems, was present on May 20 for the first light image; as part of his Institute duties he immediately began analysis of the image using optics software packages he had created. Running the various programs with different assumed values for unknown parameters within 48 hrs of the Sunday "first light" image, he was able to reproduce most features of the undesirable halos observed around the stars when he inserted fundamental optical aberrations such as coma, spherical aberration, and astigmatism into his analysis. These aberrations could only exist if the *HST* primary and secondary mirrors were either misshapen or misaligned—a very serious prognosis. He presented his findings, which indicated aberrations as a significant cause of the less-than-ideal images, at the May 22 meeting of the *HST* Project at Goddard, acknowledging that his conclusions were preliminary because he had not satisfactorily accounted for all the structure of the stellar halos. For more than a week, Burrows' interpretation was discussed but gained little traction within the Goddard meetings, mainly because independent analysis by the NASA contractor who had configured the mirrors demonstrated that improper positioning and orientation of the secondary mirror could not be ruled out as the cause of the flawed image structure. Given the strong reaction that would be forthcoming if a misshapen mirror were the cause of the faulty images, it is not surprising that Burrows' interpretation held little appeal as long as a less consequential explanation for the blurred images existed.

Further data were needed to understand better the *Hubble* image characteristics, so additional images of a more compact star cluster that provided for more stars across the detector were taken on May 31 with the secondary mirror in a different position believed to be its optimal location, and these were made available for analysis. The WF/PC instrument team, whose science with the *Hubble* depended critically on having superior image quality, took great interest in understanding the cause of the poor images so they could be improved. Members of the team became increasingly interested in—and concerned about—Burrows' software analysis suggesting serious aberration as the source of the fuzzy images. Sandra Faber, an eminent astronomer from the University of California, Santa Cruz, and member of the WF/PC team, took the lead with several of her collaborators, including J. Holtzman and T. Lauer, in trying to model the entire optical path of the telescope and WF/PC instrument. Using some of Burrows' basic software, they developed more refined packages of their own that included a complex variety of optical aberrations and mirror and instrument configurations that predicted the resulting image structure for different secondary mirror tilt and focus positions.

In order to compare their results with actual telescope images, the WF/PC team proposed that a series of images bc taken with the telescope at different secondary mirror focus positions. The structure of the deliberately out-of-focus images would provide essential information as to the cause of the apparent poor image quality of the telescope. The Hubble Project staff within NASA were very reluctant to move the secondary mirror far from proper focus, for fear that it might get stuck such that it could not be moved back to its optimal position. For two weeks, they argued against doing the test run while continuing analyses of the two available existing images were being presented to them daily.

The image analyses, still based on relatively scant data, fueled intense discussions in meetings and private corridor conversations, slowly convincing everyone associated with the Project that more data were necessary if the telescope optics were to be understood. The Project relented, and in mid-June authorized the risky run of seriously out-of-focus images, uploading to the telescope the commands for the test run to be executed. Fortunately, all went smoothly with the positioning of the secondary mirror along its focus track and the telescope accommodated by taking a series of deliberately out-of-focus images, downloading them to the Project at Goddard.

Data in hand, the instrument teams began analysis of the images. The group led by Sandy Faber continued its use of Burrows' software to generate extensive software models representing plausible causes for the poor images. They ran large numbers of computer simulations in which different values of the parameters of the telescope and instrument optics were assumed, for which the output created theoretical images that could be compared with the recently executed out-of-focus images. One of their simulations, shown in Figure 3.3, represented a major spherical aberration caused by a misshapen primary mirror; it stood out as showing remarkably good agreement with the series of actual out-of-focus images. The key to demonstrating that spherical aberration was the primary problem causing the poor images was their circular symmetry. Coma and astigmatism could be ruled out as significant aberrations because they should produce asymmetries in the images that were not observed.

As explained to me by Dr. Faber, "There were only two unknowns for each image in [our model] focus runs: focus and spherical aberration. We adopted a value of spherical [aberration] and then tweaked the focus. The goal was to match the entire sequence with the same spherical aberration AND a set of focus values that differed by the known movements of the secondary mirror between images. Hence, just two unknowns [were relevant]: absolute focus position and spherical aberration. Looking at a given image, there is a lot of degeneracy between the two as bad focus and bad spherical both make images larger. A key image to fit was therefore the radius of the smoke-ring image. That really pins down the value of spherical aberration independently."

In a memorable meeting with the Hubble Project, the thorough calculations of Sandy's group were presented by team member Jon Holtzman to the major NASA and contractor players in a convincing manner that produced stunned silence: the already orbital *Hubble Space Telescope* suffered a serious spherical aberration that could not be corrected without a fundamental re-working of its main optical system.

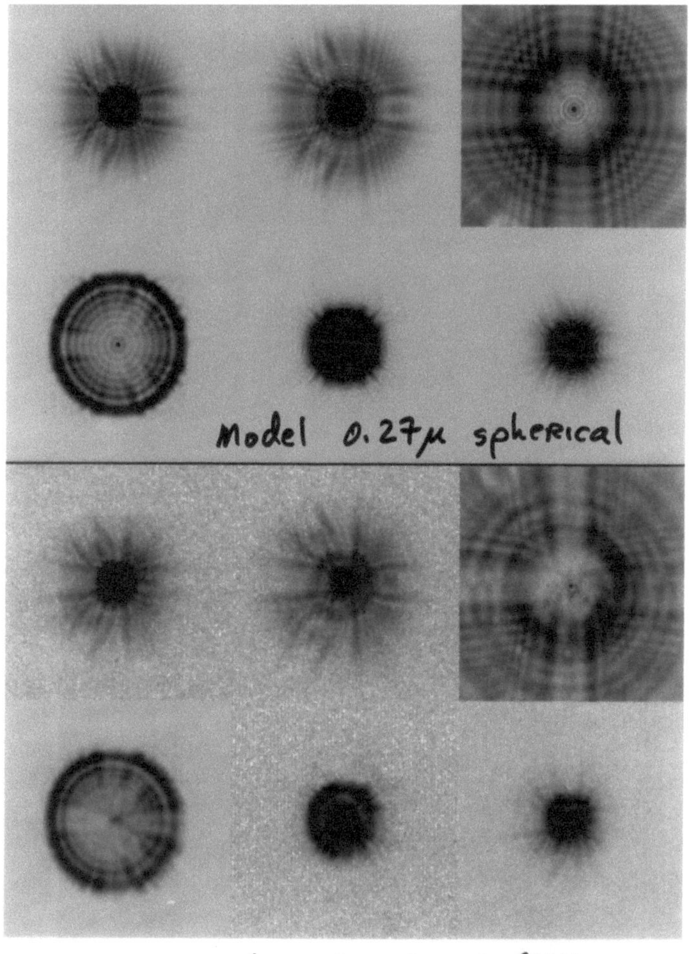

Figure 3.3. A comparison of actual images of a star taken with the WF/PC for different focus positions (bottom) vs. calculated model images expected for the same focus positions as determined by an optics ray tracing program that assumed spherical aberration (top) [credit: Sandra Faber].

The optical analysis of the out-of-focus images generated by the Faber and Burrows groups showed that the *Hubble* mirrors were misshapen—the primary mirror was flatter than it should be, by the tiny amount of 4 microns. This miniscule defect was sufficient to turn *Hubble's* intended razor-sharp images of stars into fuzzballs with surrounding halos—a debilitating diagnosis.

By the end of June, NASA advised Congress and the public that the *Hubble's* capabilities were "seriously compromised" and that its future was in doubt. Public reaction to the debacle of the multi-billion dollar Hubble Project was swift and uncompromising, as is evident from the political cartoon in Figure 3.4. Universal outrage in the print media and over the airwaves focused on NASA ineptitude and

Figure 3.4. One of the many political cartoons in the media that reflected the reaction of the public to *HST's* serious optical problems [credit: Englehart/Hartford Courant].

the high cost of doing astronomical research. In this desperate situation, the saving grace for astronomy and NASA was that the *Hubble* was already in orbit and therefore not easily consigned to an early grave. This respite allowed NASA and the astronomical community to group together quickly and formulate an ingenious plan to fix the telescope's problems. The plan could have turned out very badly but for the fact that some extremely capable individuals, who had already been associated with the *HST* Project and STScI, dedicated themselves to solving what seemed initially like an impossible situation. Ultimately, they made the difference in turning *HST* from a flawed instrument into an icon of scientific discovery.

Spherical aberration is one of several optical distortions that can occur when lenses or mirrors are shaped to image light. In a single lens reflex camera, these aberrations are corrected by combining separate lenses together to correct the aberrations. Your Nikon camera lens actually consists of a combination of distinct lenses made of different material, and shaped so as to eliminate these aberrations. In the case of imaging mirrors, the situation is simpler but the same principle applies. One can insert a corrective mirror into an optical system that has aberrations and shape the mirror to correct for the aberrations introduced by the other mirrors in the system. In this way, the finest images possible are obtained. This potential solution to *HST's* spherical aberration was quickly realized by scientists associated with the *Hubble*. The question was: in what practical way could corrective mirrors be introduced into the telescope? Congress was upset with NASA and the astronomical community for having produced, at huge cost, a telescope that didn't work well. They wanted answers as to why things went wrong, and serious suggestions for a way forward. However, it was the American public, whose taxes had seemingly been wasted, who were even more upset. They were calling for the *Hubble*—jokingly referred to as *Rubble Space Telescope* by this author's own son—to be shut down, scuttled, mothballed, or whatever one did with a decommissioned orbital boondoggle.

STScI, who were contractually responsible for carrying out the science program of the telescope, responded to this by forming a community-wide strategy committee to consider options for fixing the *Hubble*'s problems. Spherical aberration was not the only trouble. The solar arrays were bi-metallic, constructed of two metals that had different expansion rates. Whenever the telescope passed from daylight to the nighttime portion of its orbit, the arrays responded by snapping into a different shape. Like the burping of a Tupperware container, the entire telescope assembly responded by vibrating for some minutes, further degrading images already blurred by spherical aberration.

The *Hubble Space Telescope* was always an integral part of the NASA Shuttle program so that it could be serviced, with faulty components and instruments replaced by newer technology components. A path to possibly correcting all of *HST's* problems thus existed while it remained in orbit. STScI's Strategy Committee received many suggestions and proposals for restoring *HST* to its full capabilities. Given the public outcry over a crippled facility, NASA and the Strategy Committee resisted calls for *HST* to be placed in the Shuttle and returned to Earth so it could be repaired in a proper lab, knowing that if *HST* were ever returned to Earth, it would almost certainly remain grounded forever. There was no chance the political environment would allow it to be relaunched. Instead, a plan was formulated that would enable astronauts to access critical parts of the telescope while it was attached to the Shuttle and perform the swapping of components needed to restore the *Hubble* to its full capabilities.

The *Hubble* restoration plan was a collective effort that came together in piecemeal yet relatively quick fashion via meetings of the many groups (e.g., NASA, contractors, instrument teams, STScI, and technical consultants) that had worked together to create the launched telescope. The key to restoration was finding a way to insert corrective mirrors into the main optical path of the telescope. This was shown to be possible via an ingenious solution, conceived of by STScI engineer Jim Crocker, that required removing one of the instruments located at the rear of the telescope. Because the instruments were located in different positions, the corrective mirrors needed to be placed on a stem that allowed the different corrective mirrors to move in and out and swivel into different places, depending on what instrument was receiving the light from the *Hubble* mirrors. Crocker and collaborators created a clever design for such an assembly, called COSTAR (shown in Figure 3.5) that solved this problem in an efficient way that would genuinely solve the *Hubble's* aberrated images with little loss of sensitivity.

The solar array difficulties were more easily resolved by fabricating a completely new set of arrays that had already been designed for and used on commercial satellites and did not have the expansion and contraction problems experienced by those on the *Hubble*. Shortcomings had also been found in other systems that had less operational impact, either because redundant systems existed to replace them or workarounds were found to bypass them. Correction of spherical aberration remained the overriding problem plaguing *Hubble Telescope*, rendering the telescope a far less potent scientific instrument than planned—although still more useful than the best ground-based telescopes of that time.

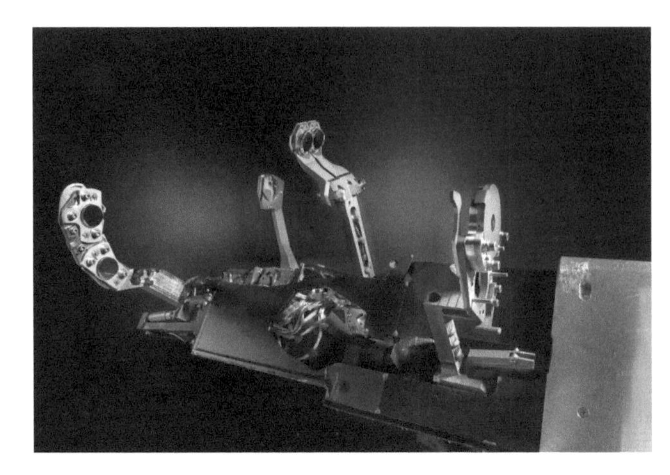

Figure 3.5. The COSTAR corrective optics assembly of movable mirrors that was inserted into *HST* during the first servicing mission to correct the *Hubble's* spherical aberration [credit: Eric Long/NASM/Smithsonian].

Within its first six months of operation, intensive analysis of every aspect of *HST's* functions had produced a clear picture of what worked as it should and what did not. Aside from the serious problem of spherical aberration and the vibration of the solar array panels following entry and exit from the sunlight portion of the orbit, the telescope was performing well. Correction of the optics and solar array problems would restore the telescope to its optimal performance, and the plan that had been crafted to solve the problems would require a successful servicing mission visit to the telescope by a team of astronauts, who would perform the delicate tasks in the hostile environment of low Earth orbit. Quite separately, those astronauts had been selected and were in training in Houston at Johnson Space Center as soon as the servicing plan was finalized.

AAS | IOP Astronomy

Hubble Deep Field and the Distant Universe

Robert Williams

Chapter 4

Hubble Servicing Missions

The *Hubble Telescope* was designed and built to be serviced in space by astronauts from the Space Shuttle. For this reason, it is located in what is referred to as *low Earth orbit*, 600 km above the Earth's surface. Because of the large mass of the telescope, 11 metric tons, this was the highest orbit the Shuttle could achieve with *HST* in its cargo bay. At this height, the telescope is above more than 99.99% of the atmosphere, yet the scant air still produces sufficient drag on the telescope to cause its orbit to decay as much as one kilometer every year. Counteracting this requires a boost by the Shuttle every servicing mission, to compensate for the increased drag and an increasingly rapid descent as the orbit decays. Now that the Shuttle has been retired, without additional orbital boosts the remaining orbital life of the telescope is on the order of 20 years before atmospheric drag brings it back down to Earth. The timescale is uncertain because it depends on the extent to which solar activity, which serves to heat the atmosphere, causes the upper atmosphere to expand such that drag increases and hastens the descent of the telescope.

Some of the more massive components of the telescope assembly, e.g., the reaction wheels and mirror, would likely survive its eventual plunge back to Earth, so NASA is committed to eventually de-orbiting the telescope into the ocean in a controlled manner. The final Shuttle servicing mission visit to the *HST*, which took place in 2009, installed a special connecting ring on the back of the telescope assembly to which a rocket propulsion package can eventually be attached in order to perform the de-orbit procedure.

Much of the equipment on the *Hubble* that involves electronics and moving parts that can fail has been installed in duplicate, e.g., six gyroscopes are provided, only three of which are needed for the precise pointing of the telescope. Components that were considered susceptible to failure were designed so they could be accessed and replaced by astronauts. Understandably, duplication or replacement is not viable for the largest components of the telescope, such as the primary and secondary mirrors or the telescope tube. They, too, are at risk of suffering damage from collisions in

doi:10.1088/978-0-7503-1756-6ch4

orbit with space junk that could render the telescope inoperable. The potential for crucial parts of the *HST* to fail is an inevitable risk that must be accepted in the hostile environment of space.

The calamitous events that produced the misshapen primary mirror and spherical aberration demonstrated that detailed plans are no guarantee against major mistakes, which may resist easy solutions. Fortunately, resourcefulness often shows up in difficult situations when there is inspired leadership present. This was the case when the first servicing mission, SM1, to the *HST* became a necessity. When NASA was presented with and agreed to the detailed plans to fix the *HST's* serious performance problems, it did what had to be done, finding the necessary funds and authorizing the repair mission. General management and oversight of the mission was assigned to Goddard Space Center. Astronaut training was to be handled by the astronauts' home base, Johnson Space Center (JSC) in Houston, and overall leadership of the critical aspects of the mission fell to two highly qualified Goddard professionals who were associated with the *HST* Project, Joe Rothenberg and Frank Cepollina. Joe was Director of the *HST* Project and Frank was Manager of Space Servicing at Goddard. They led the consortia that formulated the detailed plans for SM1, which were to correct the problems with *HST* and put it back in prime working order. No small task, but they both proved themselves to be remarkable leaders, among the most capable this author has ever worked with, who deserve great credit for managing a complex situation in a way that produced a historic moment for space astronomy and NASA.

Servicing missions had been foreseen for the *Hubble* throughout its nominal lifetime, assumed at launch to be on the order of a decade. Initially, it was assumed that servicing missions would be programmed every 3–5 years, but there was no set schedule. Rather, it was understood that system failures and upgrades based on the best new technologies would be necessary periodically to keep the telescope operating at the best level possible. The cost of a servicing mission, on the order of $1 billion when all factors such as the Shuttle launch and instruments were included, served as a natural deterrent, so good reasons were needed to justify a servicing mission. An *HST* crippled by spherical aberration qualified in spades, so NASA was compelled to develop a mission as soon as it was presented a credible plan that had a reasonable chance of being carried out safely and successfully.

As already mentioned, a very clever fix for the spherical aberration had been suggested. This solution involved the replacement of one of the *Hubble's* initial instruments, the high speed photometer, by the COSTAR suite of corrective mirrors. It would also be necessary for these astronauts to swap the original solar arrays for newer ones, which were already in use on other satellites, that did not change shape when passing from Sun to shadow. As soon as funding was made available by NASA for the mission, even before it had formally put the Shuttle on its launch manifest, the agency had selected the team of astronauts for the proposed mission to begin their rigorous training.

Shuttle servicing missions to the *Hubble* involved a complex web of activity carried out by different teams with separate tasks that required close collaboration. The HST Project at Goddard was in charge of the mission's interface with the telescope. Kennedy Space Center at Cape Canaveral, working with JSC in Houston, had primary responsibility for the readiness of the booster rockets and Shuttle. Goddard worked with the telescope instrument teams and managed the data-taking chain, including telescope pointing, gyros, batteries, computers, and communication antennae. All three NASA centers collaborated on the astronaut training at JSC. STScI worked closely with the Project on all matters that might impact the science from the telescope. To say that full servicing mission preparation resembled a military campaign is no overstatement.

Apart from the critical nature of the astronaut training, one of the core elements of the preparation for *Hubble* servicing involved hardware engineering and system integration. For SM1, the instrument teams, Ball Brothers Co. engineers (who built several of *Hubble's* primary instruments), and *HST* Project staff at Goddard worked together feverishly to integrate all the activities. Within the short span of 2–3 months, they had worked up a plan to install the corrective optics. Remarkably, already by late summer 1990, an optical assembly had been designed that was build-ready. Solutions for the other problems with the telescope, including the solar arrays, were more straightforward, such that the engineering aspects of the first repair mission to the *Hubble* already looked good by the end of 1990, only eight months after the telescope launch.

Apart from the detailed hardware design, build, and testing component of servicing missions that have kept the *HST* operational for so many years, it is certain that the remarkable performance of the astronauts in space will endure as one of the most important legacies of the Shuttle missions. The *Hubble* missions were the most coveted of all the astronaut opportunities, insofar as none of the other missions required the complex work that was required to keep the *HST* at maximum performance. The selection of the astronauts was an internal NASA matter handled within the Astronaut Office; it normally required an intense 1–2 year commitment to mission training for the astronauts tasked with EVA (extra-vehicular activity) work inside and outside the telescope assembly.

Training began as soon as it was determined what equipment needed to be repaired or swapped out with replacements. Ferrying large equipment between the Shuttle bay and the telescope and then gently inserting it into the telescope required a series of very sensitive maneuvers in which any wayward movement could cause a collision between new and in-place equipment that would damage both. Thus, one of the essential training exercises for *Hubble* servicing missions was the rehearsal of all equipment movement and installation in weightlessness, including manipulation of delicate and small instrument components.

How can weightless conditions be created in a situation where astronauts are handling large equipment? The answer is to train underwater with flotation gear that is set to balance the effects of gravity for both humans and equipment. The huge Neutral Buoyancy Lab tank at Johnson SFC was constructed for just this purpose, so astronauts could don pressurized suits and practice the movement and maneuvering of large equipment underwater in a weight-free environment (Figure 4.1). Many hours were spent in the tank before each servicing mission, doing repeated practice runs where the two-astronaut EVA teams coordinated their activity so they

Figure 4.1. Astronauts Jeff Hoffman and Story Musgrave underwater in Johnson SFC's Neutral Buoyancy Tank, practicing insertion of the WFPC2 instrument into the *Hubble Telescope* instrument bay [credit: NASA].

could perform every task in a way that was timely and did no harm to any telescope component.

The astronauts needed to be intimately familiar with every screw that required loosening, every wire connector to be disengaged, what nearby components could not be touched for fear of damage, the lighting required to perform necessary tasks, and a thousand other details that might require sudden decision-making during each task. As an example, in one epic servicing mission EVA, an electronic circuit board needed to be replaced in one of the two most utilized *Hubble* instruments, the STIS spectrograph, which had not been designed such that its circuit boards could be serviced in orbit. The importance of repairing this instrument led to a very difficult, awkward series of maneuvers and tasks by the astronauts. Astronaut Mike Massimino was required to loosen more than 100 screws, capture and collect them in a container while the circuit board was swapped out, and then finally reinsert the screws to close up the instrument again. The task required a complicated choreography, including the unexpected need to break off an overlying hand rail, and it was successful in restoring the vital STIS instrument back into use.

Not all of the unexpected contingencies have occurred while the astronauts were in orbit. Some took place in training. On one occasion, SM1 astronaut Story Musgrave was tasked to loosen a recalcitrant screw that was frozen cold as part of a training thermal test. Having some difficulty with it, Story decided—against procedure protocol—to remove his glove in order to get a better grip. His efforts instantly resulted in fingers that became seriously frostbitten at a critical time very close to the SM1 launch. The servicing mission leadership were understandably agitated (to put it in the most diplomatic language) by Story's risky mishap because it appeared likely that he might have to be replaced by another astronaut who lacked training for the SM1 tasks. Desperate to do whatever it took to get Story's fingers healed, NASA flew him to Alaska on an urgent trip for a medical consultation with one of the world's foremost experts on frostbite. Fortunately, whatever therapy was applied worked; Story's fingers did heal in time for his SM1 activities, but not before NASA managers lost (more than) sleep over the incident.

Apart from familiarizing themselves with every detail of whatever equipment they planned to be working on during their EVAs, the astronauts needed to be comfortable with accommodating to the weightlessness of Earth orbit. This was especially crucial for the one astronaut who was the *free floater* of the two-person tandem that performed *Hubble* EVAs. Each EVA team had one astronaut with feet stabilized by being clamped onto the long-armed Remote Manipulation System—the so-called Canada arm. The other astronaut would be free-floating, tethered to the Shuttle with a loose cable so s/he could not float away. All movements of the free-floater took place via their hand-walking around the telescope, using the numerous yellow hand rails positioned on the outside of the telescope tube assembly. This is clearly seen in Figure 4.2 where astronaut John Grunsfeld, known as the *Hubble* Repairman due to his resourcefulness while participating in three of the five *HST* servicing missions, is seen holding onto one of the telescope hand rails while he assists in the refurbishment of *HST* hardware during the first EVA of the final servicing mission in 2009.

The *Hubble* servicing mission astronauts recount many interesting stories of their experiences living and working for ten days in a weightless environment. The necessities of eating, sleeping, organizing equipment, and performing daily 8 hr EVAs left little time for any extraneous activity. The schedules of the entire crew were choreographed in detail, accounting for every 15 min increment of the 24 hr day. Not much opportunity existed for leisurely gazing through the Shuttle windows at the glorious Earth drifting by. Individual astronauts had different reactions to the unusual conditions they found themselves in when up in orbit, and they found various ways to acclimate to performing necessary functions, e.g., sleeping in tethered bags attached to the Shuttle walls. Story Musgrave, who was not only a free floater on an EVA of his SM1 mission but also a free spirit, enjoyed sleeping untethered, floating in the Shuttle airlock. Various astronauts also shared a not-uncommon tendency to have strangely vivid, wild dreams while in orbit.

There was another sensation that astronauts experienced when they were free floating on EVA repair work around the *HST* assembly: fear of falling. Despite the fact that the free-floating astronaut was tethered to the Shuttle by a flexible cable and occupied the same orbit as the telescope, so there was no reason they should drift away, they did experience the strong sensation of falling. The instinctive response to grab and maintain a tight grip onto whichever handrail was nearby when working

Figure 4.2. John Grunsfeld, the free floater for the first EVA on the *HST*'s final servicing mission in 2009, is assisting his partner Drew Feustel in removing the WFPC2 instrument that was to be replaced by the infrared sensitive WFC3 camera. The cable securing Grunsfeld to the Shuttle can be seen extending downward from his waist belt [credit: NASA].

on the telescope was unavoidable. I remember very well giving a lecture with SM1 and SM3 *Hubble* astronaut Claude Nicollier, when I mentioned with naïve puzzlement the fact that free-floater astronauts were unable to overcome their sense of falling. Claude, having been one of the space-walking astronauts on the third *Hubble* servicing mission, immediately interjected, "But Bob, we ARE falling! The feeling is overwhelming."

On one particularly dire occasion, a free floater found himself continually grasping the yellow hand rails around the telescope during a long EVA, to the point where his arms started cramping up badly and he was only marginally able to complete his tasks. The situation become very dicey because the scheduled instrument swap out had not yet been completed and the EVA team was near the end of the consumables they needed to remain outside the Shuttle. The NASA controllers on the ground at Johnson SFC came very close to making the decision to end the EVA for astronaut safety concerns. The consequences of failing to install that particular instrument in the telescope would have been enormous because it was crucial for important scientific observations. Fortunately, the astronauts on that particular EVA were able to complete the task successfully just before they were forced back into the Shuttle.

There were five separate Shuttle servicing missions to the *Hubble* in the interval of 1993–2009. The first mission received a huge amount of public attention because of its importance in working to correct the spherical aberration that plagued the telescope. Images of the astronauts at work in space—such as the one shown in Figure 4.3, where Jeff Hoffman and Story Musgrave are performing the WF/PC instrument change out that they had practiced so diligently in the Johnson SFC buoyancy tank—were splashed across the front pages of newspapers around the world. However, each of the subsequent missions was equally essential to maintain the *HST* in good working order. All of the SMs had some common themes and processes. For example, most of the missions had two two-astronaut teams to do the EVAs, alternating with each other. A fifth astronaut crew member remained in the Shuttle, tasked with the responsibility of controlling the movement of the Canada arm with its attached astronaut. The mission commander remained in the Shuttle to control its functions, especially its maneuvering. The EVA spacewalks were designed to have a duration of roughly eight hours, representing the maximum time that the

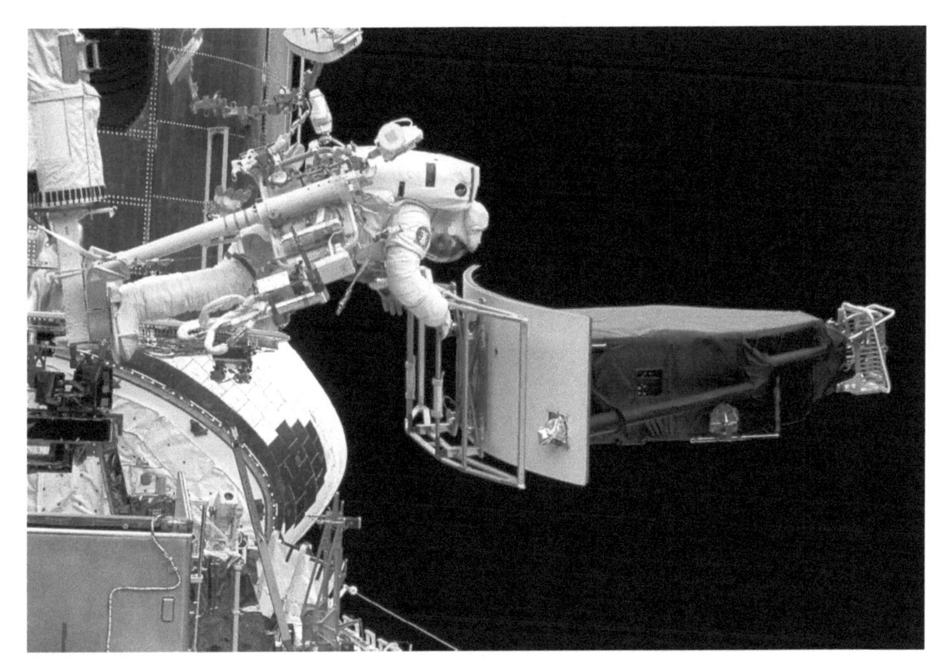

Figure 4.3. Astronaut Jeff Hoffman, anchored to the Canada arm while removing the original WF/PC from the *HST* during the 1993 first servicing mission. After storing the instrument in the Shuttle bay, Jeff installed the new and improved WFPC2 instrument in the telescope [credit: NASA].

combined space suit + backpack ensemble could perform its functions, e.g., provide oxygen and prevent water vapor condensation from clouding the visor.

The work and sleep schedules of the astronauts on most Shuttle missions with EVAs were tightly constrained by a distinct feature of the Earth's Van Allen radiation belts called the South Atlantic Anomaly (SAA). The Van Allen belts have a structure that is determined by the Earth's magnetic field, and they present a dangerous environment for humans and electronic equipment. The belts contain highly energetic atomic particles, cosmic rays, that are captured from the solar wind by the Earth's magnetic field and possess harmful, DNA-damaging energy. The major portions of the Van Allen belts occur at altitudes exceeding 1000 km above the Earth's surface, due to the Earth's atmosphere absorbing the energetic particles and serving as a protective shield below this height. However, the structure of the Earth's core creates an isolated component of the Earth's magnetic field that dips down and pierces through the Earth's upper atmosphere, off the east coast of Brazil, allowing a region of intense Van Allen cosmic rays to extend down to within 200 km of the Earth's surface near that region of the Atlantic Ocean.

For humans on the Earth's surface, the SAA presents no threat to health; below the SAA, the atmosphere still effectively shields anyone on the ground. For astronauts in orbit above 200 km, however, the cosmic rays are a serious hazard because they can easily penetrate the body and cause damaging genetic mutations. The orbit of the *Hubble* is fixed in space relative to the Sun–Earth system, so as the Earth rotates in its 24 hr period, the Shuttle's 96 min orbital period carries it through significant parts of the SAA during at least 4 of its 15 orbits each day. The astronauts must be inside the Shuttle during those passages so its metallic structure can protect them from the majority of the high-intensity Van Allen belt cosmic rays. EVAs take place only during the non-SAA impacted orbits—a constraint that determines the basic servicing timeline of astronauts in orbit. Venture outside of the Shuttle in only your space suit while passing through the SAA and you are putting your future health at serious risk from radiation damage.

The SAA poses a risk to more than humans. Sensitive electronic equipment is susceptible to circuit board damage from the large flux of high-energy atomic particles because they are charged particles themselves and constitute a form of electric current. A large influx of cosmic ray particles can burn out an integrated circuit. As a precaution, *HST* does not take observations while the telescope is in the SAA, and some of its most sensitive electronic equipment is powered down during SAA passages. Satellites that have not taken similar cautionary action have shown high equipment failure rates during SAA passage. Protection afforded by the Shuttle during SAA passage is not complete; a small fraction of cosmic rays do pass through the bodies of the astronauts. Cosmic ray particles are known to activate the retina when they strike it, producing a yellow flash. Some astronauts eyes have a special sensitivity to cosmic rays and they report that they are awakened from sleep when the Shuttle passes through the central, most intense part of the SAA. For them, the yellow flashes are sufficiently bright and numerous that they disrupt good sleep.

Even outside the SAA, there is never a total absence of cosmic rays passing through whatever happens to be orbiting above the Earth's atmosphere. There is a small flux of cosmic rays at all times and they leave their marks on the CCD detectors of the *Hubble* in the same way that light does. If one looks at an unprocessed image of any celestial object taken with *HST* with an exposure longer than 30 min, it will show an impressive number of streaks and blotches that are the product of cosmic ray *hits* on the detector. Figure 4.4 shows a typical examine of such an exposure that must be cleaned of the *noise* features produced by the cosmic rays before useful astronomical analysis can begin. Such cosmic ray hits constituted one of the factors that determined the faintest galaxies detected in the Hubble Deep Field image after digitally adding up all of the individual exposures.

All five servicing missions to *HST* served important purposes in keeping the telescope functioning as a unique facility, and they demonstrated the ability of humans to perform complicated work in space. It goes without saying that they also served as textbook examples of how risk-taking, with its inherent danger of failure, can produce results that push back frontiers. The rescue and renovation of *HST* by astronaut servicing has played a key role in the debate over the value of serviceable space facilities, which does add large expense to a project, as opposed to unserviceable facilities that are not designed to be visited and serviced by astronauts and which are therefore cheaper to produce. Both points of view have validity, however, so the debate cannot be resolved by one answer that is applicable to all situations.

The development of robotics, whereby machines can perform many of the same functions as an astronaut, is likely to have an important impact on this issue. Without a manned Shuttle to ferry astronauts to attach needed rocket boosters to the telescope, the eventual *HST* de-orbit will likely need to be managed robotically. Serious studies are already underway for the development of a robotics package that can visit *HST* in order to attach a rocket propulsion system to eventually de-orbit the telescope (Figure 4.5). It is likely to be the most practical solution for a safe de-

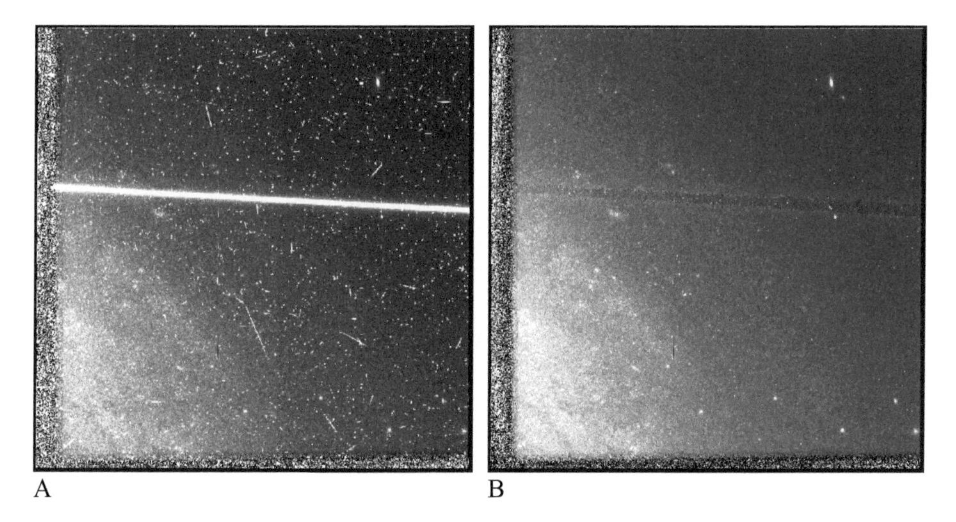

Figure 4.4. Two versions of the same *Hubble* exposure. The left shows a raw frame with numerous cosmic ray hits and a bright satellite streak through the field. The right shows the same exposure but with imperfections removed by a software package [credit: Michael Richmond/RIT].

Figure 4.5. The design of a NASA robotics module under consideration for testing in the Goddard robotics lab. Such a module could eventually be launched to service the *HST* by attaching a propulsion module that would safely de-orbit the telescope into the ocean [credit: NASA].

orbit of the telescope—which many of us wish instead could be de-orbited intact and placed in the NASA Air and Space Museum as a tribute to mankind's efforts to better understand the universe.

Focusing on the seminal first servicing mission, SM1, it was one of those endeavors where everything worked as it was supposed to. There was a great deal of tension about how things would turn out, especially given the universal criticism that rained down on NASA and the astronomical community over the spherical aberration. In fact, the professionals within NASA at Goddard and the collaborative groups at STScI, Lockheed Martin, and Ball Brothers steadfastly performed every aspect of the work had been prescribed in support of the mission. Once in orbit with the *HST* anchored to the Shuttle, the two astronaut teams performed the EVA tasks. The team of Story Musgrave and Jeff Hoffman installed new gyroscopes and assorted electrical components on their EVA1. The following day, during EVA2, the team of Kathryn Thornton and Tom Akers extracted the original solar arrays, one of which could not be rolled up as planned, and installed their replacements. The original array that could not be rolled up produced a brief hiccup in plans because it could not be stored safely in the Shuttle payload bay. There was no other option than to carefully release the array in space so that it would not collide with the telescope. A memorable moment found Dr. Thornton releasing the array from her gloved hands, with her feet anchored on the Canada arm as it was slowly retracted back toward the Shuttle, leaving the disabled panel in its own orbit (Figure 4.6). The array's orbit slowly decayed for more than a decade before the panel eventually vaporized in the upper atmosphere.

The crucial EVAs related to fixing the spherical aberration of the telescope took place over the following days. In EVA3, Hoffman and Musgrave replaced the original Wide Field/Planetary Camera (WF/PC) with WFPC2, which was very similar to its predecessor but had the additional optics that corrected the spherical aberration of the *Hubble's* main mirrors. This camera would be the workhorse instrument for the telescope until a subsequent servicing mission installed a more

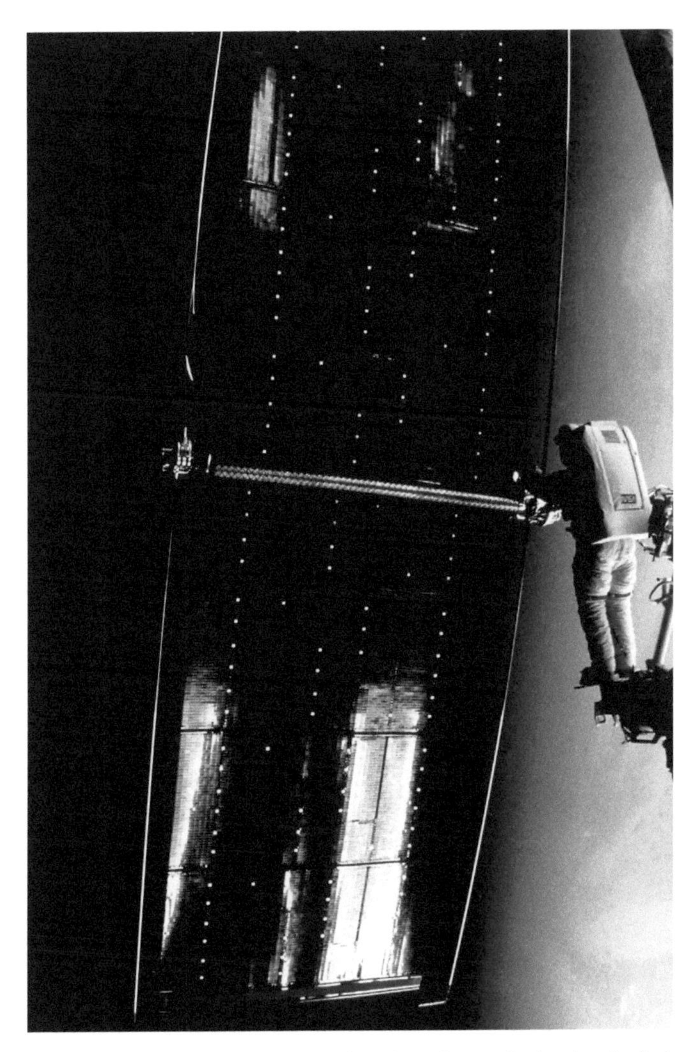

Figure 4.6. Astronaut Kathryn Thornton at the end of the Canada arm, jettisoning the faulty original *Hubble* solar array. The Earth appears in the background [credit: NASA].

technologically advanced camera. In EVA4, Dr. Thornton and Col. Akers removed the High-Speed Photometer instrument and inserted in its place the COSTAR corrective optics module that served to redirect and rectify the spherically aberrated *Hubble* optical beam for the remaining *Hubble* instruments. They also installed a new coprocessor in the main *HST* computer to enhance its speed and memory.

Having finished the planned spacewalks and checked out the telescope to make sure it was in prime working order, the astronauts relaxed, preparing for their return. On the ground, NASA was ecstatic. Although the first images to be taken with the refurbished telescope were not planned for another week, the anticipation was that the *Hubble* had been restored to its original capabilities. The post-SM "first light" images were taken on 1993 December 18, six days after the Shuttle Endeavor had landed. Unlike the original first light *Hubble* images taken following launch in 1990 April, which showed the serious spherical aberration, the post-SM event was not advertised to the media and was attended by only a handful of knowledgeable *HST* Project and STScI staff. The initial images taken with the newly installed WFPC2 appeared quite good. After a quick analysis by Chris Burrows and others who had analyzed the initial *HST* images with spherical aberration, the new images were declared to be ideal—as good as the *Hubble* optical system could do.

As is often said, success has many fathers—of whom Lyman Spitzer could claim some primacy for the *HST*. But the *Hubble Space Telescope* had only one true mother: NASA's aforementioned Nancy Roman, whose vision in a NASA leadership position shaped U.S. space astronomy for decades. Just as importantly, the telescope had one truly historic godmother—Senator Barbara Mikulski. She was

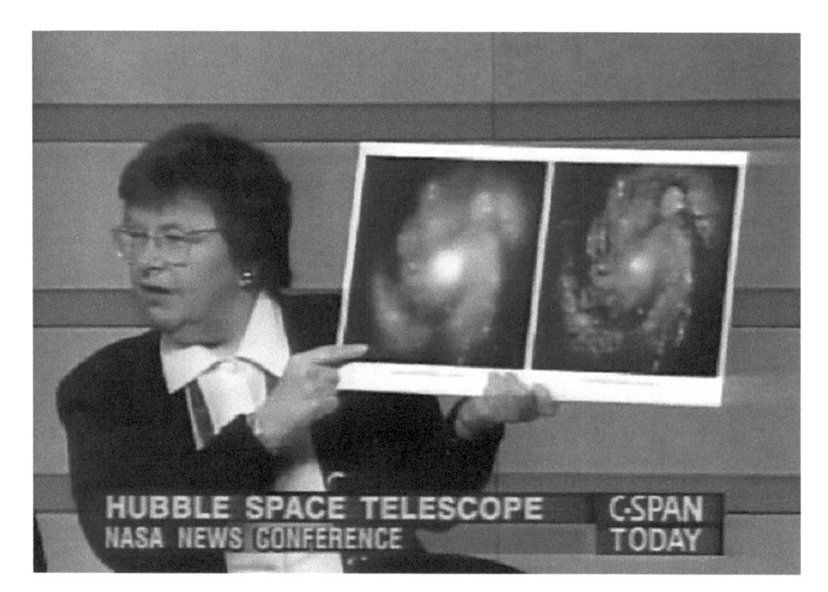

Figure 4.7. Sen. Barbara Mikulski announcing the successful refurbishment of the *Hubble* before the national media. She is holding the now-classic images of the spiral galaxy Messier 100 that were taken by the *HST* with the WF/PC (left) and WFPC2 (right) instruments before and after spherical aberration had been corrected [credit: Copyright C-SPAN].

essential in constantly navigating *HST* funding through Congressional procedures. She was also an important presence at the STScI, advocating for a strong education and outreach program based on *HST* and other NASA mission discoveries. As soon as the success of the first servicing mission's major goals had been confirmed, it was Senator Mikulski who presided over the press conference at Goddard to announce to the world that "The trouble with *Hubble* is over!" [the Senator's exact words] (Figure 4.7). For those of us associated with the telescope, it was cause for relief and joy, given all that was at stake had SM1 not gone as well as planned. The *Hubble* was now set to carry out the observations that astronomers had been preparing for all along. What remained to be seen was just what those discoveries would be.

Robert Williams

Chapter 5

Science Coffee and Distant Galaxies

The *HST* was confirmed to be in excellent operating order in mid-January of 1994, only weeks after the first servicing mission had been completed, and the Space Telescope Science Institute was prepared to go full bore with the science program it had crafted along with the astronomical community. All of the research observations had been proposed in anticipation of the restored telescope, and they had been evaluated and approved for execution by the annual peer review panels of international scientists even before the SM1 launch. A few of the programs were predominantly outreach-related, carried out in collaboration with NASA headquarters, ESA, and Goddard, and provided attractive, educational images of interesting astronomical objects. A substantial fraction of the proposals were updates of earlier proposals that had been approved for observation three years previously, before the *HST's* spherical aberration made them untenable. The astronomical community was in high anticipation of what new results would emerge from the first months of operation of the resurrected *Hubble* telescope.

The STScI, situated on the Johns Hopkins University campus, was the focus of much of the scientific excitement because of the Institute's responsibility in scheduling observations and calibrating and characterizing the instruments. It made sure the data had been received correctly and were ready to have instrument anomalies removed before being disseminated to astronomers. Raw data from the telescope inevitably contain instrument signatures that are introduced into the data by the instrument and not related to the radiation from the celestial sources being observed. Consisting of cosmic ray hits on the detector pixels and electronic noise, these need to be removed from the data before the images can be used for analysis. Data quality management is one of the important responsibilities of the Institute of essential benefit to the user community.

HST data, which include both science and engineering files that monitor the proper functioning of the telescope, are downloaded to STScI multiple times daily via a communications link that begins with data uploads from the *HST* to satellites

doi:10.1088/978-0-7503-1756-6ch5

in higher, geosynchronous orbits. The data are then downloaded to Goddard SFC in Greenbelt, MD, who microwave them to the Institute. The Institute maintains constant contact with researchers who observe with the *HST* so they are kept abreast of relevant information needed for their use of the data. All data, raw and corrected for signatures, are placed in the large data archive maintained at the Institute and a few selected mirror sites at various locations around the world.

For many years at the Institute, there has been a tradition on workday mornings of an informal 10:30 am science coffee held in the library, where the scientific and technical staff congregate to chat, socialize, and discuss the research they and their colleagues are doing. The status of the telescope and instruments, fundamental to the integrity of the data, have been of topical interest, as has information about notable research results that the scientists have heard about from within their own network of colleagues. Conversations range all over the map, in terms of interesting topics.

Immediately after the successful SM1 mission, there was particular interest at morning coffee in the upcoming collision of the recently discovered comet, Shoemaker–Levy (S–L) 9, that had been captured by the planet Jupiter and broken apart by strong gravitational tidal forces that Jupiter exerted on the comet. The train of S–L 9 fragments was headed for a collision with the planet in July, and they were expected to be vaporized in spectacular fashion as they entered Jupiter's atmosphere. Calculations had shown that effects of the collisions of the individual fragments on Jupiter would likely be visible with the *HST* even though the impact point had been determined to occur on the far side of the planet. Remembering the historical impact of asteroids on the Earth that have produced large craters now largely eroded away, interest in the S–L 9 Jupiter collision among astronomers was high. It spawned a dedicated program focused on the collision where two full weeks of *Hubble* telescope time were devoted entirely to observing the cometary collision and its aftermath. That program produced significant scientific results on Jupiter's atmosphere and constituted some of the first notable observations of the refurbished *Hubble*. Iconic pictures of the event were provided to the entire world, representing a cataclysmic event that was no doubt similar to the impact of the large asteroid with Earth 65 million years ago that resulted in the extinction of the dinosaurs (Figure 5.1).

Morning science coffee frequently divided into small groups that were interested in specific topics. Among the various conversational groups, there were those of us whose interest was directed toward the study of distant galaxies in the early universe. Aside from myself, most of that group of five or six scientists were young, only recently past having written their doctoral theses and newly arrived at the Institute. They were recruited to the Institute as researchers in short-term positions. They were academia's ubiquitous itinerant post doctoral research associates on three-year contracts, lacking a commitment of employment at the Institute beyond their fixed-term contracts. They had no institutional responsibilities so they could dedicate full time to research.

The focus of the recently arrived young scientists was to produce significant research results that would stand them in good stead in the marketplace for open

Figure 5.1. *Hubble* images of the disrupted comet Shoemaker–Levy 9 taken before its collision with the planet Jupiter. The comet image is shown superposed with the images of Jupiter. The two images of Jupiter show sites of the comet impact two hours and three days after impact, respectively. The dark impact spots are the size of the Earth [credit: NASA].

academic tenure-track positions. Deep down, one quality they had in common was a driving curiosity to understand something of importance on the grandest of scales: the structure and evolution of the universe. Having a facility like the *HST* as a tool—one whose characteristics they understood better than most other astronomers—was for them the opportunity of a lifetime. It was this that had attracted Harry Ferguson, Mark Dickinson, Andy Fruchter, and Mauro Giavalisco to take positions at the STScI, where they could be as close as possible to all the action coming out of the *Hubble Space Telescope*. They were joined in our coffee group by Marc Postman, who already had a more senior position on the tenure-track (Figure 5.2).

The enthusiasm of astronomers to study the evolution of galaxies with the *HST* was understandable in view of the way in which cosmology had flourished in previous decades from research efforts with the largest ground-based telescopes. From Georges Lemaitre's and Edwin Hubble's initial discovery in the 1920s that the motions of galaxies revealed a uniformly expanding universe, it became possible for astronomers to piece together the evolution of galaxies in the universe by measuring the recession speed of galaxies. From those velocities, they could determine the distances and thus the epochs at which the galaxies were being observed. Extragalactic astronomy became a darling of the astronomical community, consuming large portions of telescope time during what astronomers call monthly *dark time*. Because distant galaxies are faint, they must necessarily be observed when

Figure 5.2. A discussion among HDF team members at morning science coffee in the Institute library, including Harry Ferguson and Mark Dickinson, two of the driving forces behind the Hubble Deep Field Project [credit: STScI].

moonlight is not illuminating the night sky and limiting the faintness of objects that can be detected.

Dark time is the interval of 4–5 days on either side of the new Moon each month, when the sky is not brightened by moonlight. It is the most coveted time for astronomers trying to detect thc faintest objects, and because it occurs for less than ten days each month, the fraction of nights that ground-based telescopes can devote to the study of distant galaxies is limited. One of the virtues of the *Hubble* is that its position above the atmosphere essentially eliminates the problem of sky brightness. Not surprisingly, the favorable location of the *HST* in space causes it to be a coveted tool for observations of distant galaxies. However, at the time of its launch, the prospect of having the 2.4 m *Hubble* telescope make significant advances in detecting distant galaxies beyond what much larger telescopes on the ground were capable of achieving, such as the Palomar 200 inch (5 m) telescope or Keck 10 m telescope in Hawaii, was in some doubt.

In 1990 April, the very month that the *Hubble* was launched, a noteworthy article appeared in the prestigious journal *Science*, authored by Princeton's John Bahcall, the leading advocate for the *HST* in the decade prior to launch, along two of his collaborators, Raja Guhathakurta and Don Schneider. The three scientists published a thoughtful, well-researched manuscript titled *What the Longest Exposures from the Hubble Space Telescope will Reveal*, assessing the interesting results that astronomers could expect from the *HST*. Numerous areas of astrophysics were touched upon in the article, including their analysis of the ability of the *Hubble* to resolve small spatial detail in distant objects. While affirming that the *HST* "... will reveal the shapes, sizes, and content of previously unresolved galaxies," they cautioned, "We do not expect *HST* to reveal a new population of galaxies."

The careful projection of what the *Hubble* might see in looking out to the largest distances was well-researched and founded. Bahcall et al.'s caution about "a new population of galaxies" was possibly influenced by a much debated paper that had been published in the *Astrophysical Journal* (ApJ) in 1965 by Allan Sandage, entitled *The Existence of a Major New Constituent of the Universe: the Quasistellar Galaxies*. This paper, notable partly because its ApJ publication date in the May 1 issue actually *preceded* its true date of receipt on May 15 in the journal offices, on account of its enthusiastic reception by then ApJ editor S. Chandrasekhar, did advocate a new type of galaxy having attributes similar to those of quasars. However, Sandage's interpretation of these galaxies as new, distinct types of objects that were fundamentally different from known galaxies was never accepted by the research community.

All objects appear progressively fainter with increasing distance, so they steadily fade into the background of light that is caused by different sources. For the *HST*, the diffuse background light comes from: (1) the small fraction of the Earth's atmosphere that extends above the orbit of the telescope, (2) the radiation from hot gas that exists between stars in the Milky Way and between other galaxies, (3) electronic noise within the CCD detectors of the *Hubble's* instruments, and (4) very faint galaxies, below the limits of individual detection, whose overlapping images produce a faint background blur of light. All of these factors combine with the intrinsic brightness of a galaxy to determine the distance at which the galaxy becomes too faint to be detected, its brightness lost in the dim glow of the background.

At the time of the *HST's* launch, the greatest distance for which any large ground-based telescope had been able to detect a galaxy was roughly five billion light years, and this limit was hard to surpass because of cosmological effects associated with the expanding universe. The cosmological effects occur due to properties explained by Einstein's Theory of Relativity that apply to objects having large velocities, and therefore distances, with respect to each other. In the expanding universe, the farther away a galaxy is from the Milky Way, the greater will be its speed of recession. When that speed approaches a certain fraction of the speed of light, the time and wavelength intervals that are observed from the distant galaxy become so stretched that the amount of light intercepted by *Hubble* instruments from the receding galaxy very rapidly becomes much smaller. These effects act independently of and in addition to increasing distance, which further dims the radiation of very distant objects. They become increasingly important at distances greater than around five billion light years, significantly limiting the ability of telescopes to make detections of normal galaxies beyond this distance. The careful analysis of Bahcall and colleagues in their *Science* article demonstrated that the *HST's* more modest aperture, as compared to much larger ground-based telescopes, would likely limit its ability to probe the universe of galaxies to distances significantly greater than those already achieved from telescopes on the ground.

The importance of observing distant galaxies lies in the fact that looking out in space is equivalent to looking into the past because of the travel time required for light to arrive at our telescopes from distant galaxies. We see distant stars and galaxies as they were when the light left them long ago. Astronomy is the only science that can truly observe the past directly. The rate at which the universe is expanding has now been measured sufficiently accurately that astronomers can determine just how far back in time they are observing a particular galaxy by measuring its recession speed from its spectrum. Projecting the expanding universe back in time leads to an epoch when the observable universe must have been in an unimaginably more compact, highly energetic, dense state. All evidence points to a beginning creation moment of the present universe 13.7 billion years ago, an event we call the Big Bang. In wondering how it could be possible to have such unimaginably high densities of material at the time of the Big Bang, it is important to remember that, as enunciated by Einstein, mass and energy are equivalent. Therefore, at the extreme conditions that pertained in the earliest moments of the universe, there are credible theories that suggest that it was energy that was present at the moment of the Big Bang, and mass, in the form of fundamental particles, was created momentarily *after* the Big Bang, from its vast store of energy.

If only energy may have been present at the moment of the Big Bang and particles having mass formed instantly thereafter from the energy, what preceded the Big Bang? Where did the prodigious energy come from? Scientists do not know and can only hypothesize because science addresses only those situations for which facts are available. We have not yet succeeded in directly obtaining or inferring information about the state of the universe before the Big Bang. Cosmologists have given attention to this situation and a number of hypotheses have been proposed that are consistent with our incomplete understanding of the physics that might apply to the extreme conditions of the Big Bang. The fact remains, if we could look outward sufficiently far to see objects or structure at a distance of 13.7 billion light years, we would be seeing the universe at the time of the Big Bang. A schematic representation of looking outward in space and therefore back in time is depicted in Figure 5.3, along with the cosmic timeline.

Can astronomers truly see back to the time of creation? Not quite. The unusual conditions of the early universe are characterized by an extremely hot fog of gas that has prevented scientists from seeing as far out in space as 13.7 billion light years. However, we have come close—but not with the *Hubble Space Telescope* or any of the large telescopes on the ground. Three satellites that were very sensitive to the detection of long wavelength sub-millimeter microwave radiation scanned the sky for years to detect faint, diffuse radiation that was produced by the very hot material that comprised the universe shortly after the Big Bang. This radiation was emitted at very high temperatures and energies, but has been shifted by the expansion of the universe into the longer wavelength portion of the spectrum, beyond the visible wavelengths to which most ground-based telescopes are sensitive. The small *COBE* (*COsmic Background Explorer*), *WMAP* (*Wilkinson Microwave Anisotropy Probe*), and Planck satellites have detected this faint radiation, whose origin is the hot gas near the time of the creation event of the present universe. The very nature of this cosmic background radiation, which has remarkably uniform intensity in every direction in space, indicates that the universe was very homogeneous, i.e., a uniform soup of matter without structure other than at very minute levels, at the time the

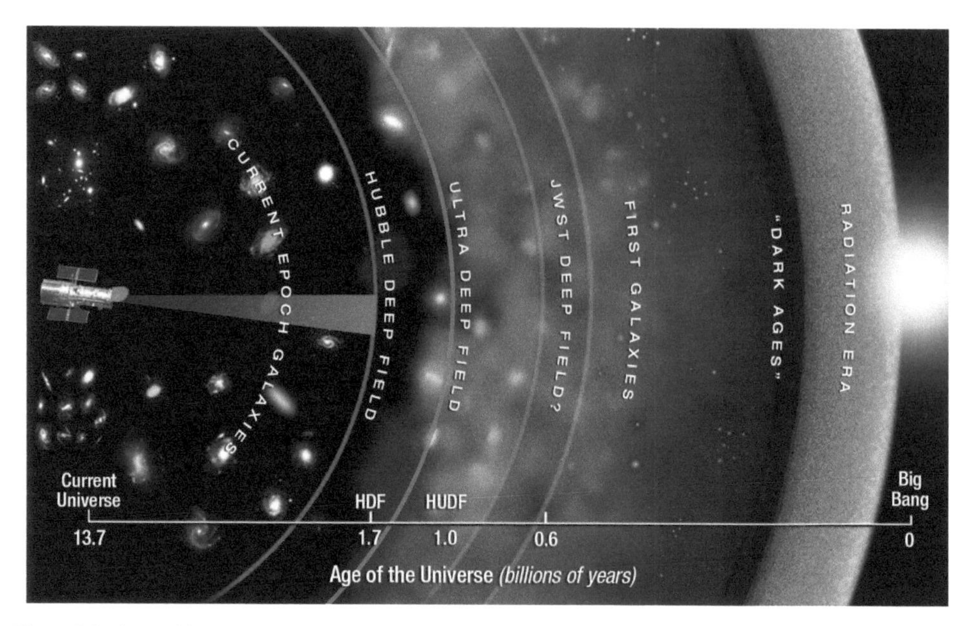

Figure 5.3. A graphic representation of the *HST* looking out into the cosmos and back in time toward the hot Big Bang. The HDF detected galaxies after they had begun to form in the universe, once its expansion had produced sufficient cooling. Successor deep fields taken with the *HST*, such as the Ultra Deep Field, probed further back in time toward the "fog" of hot formless gas that permeated the earliest universe shortly after the Big Bang [credit: STScI/A. Feild].

radiation was created. The map of this uniform radiation field was presented in Figure 1.1 of the first chapter.

The hot cosmic soup that was created by the Big Bang is its own shroud, preventing radiation produced within 400,000 years of the universe's origin from escaping for us to observe. Thus, scientists have been successful in seeing back to within a few hundred millennia of the Big Bang, but not earlier. However, observing back to a time so close to the Big Bang creation event does allow astronomers to directly observe the history of the universe, from the time when structure—stars, planets, galaxies—first began to form 13 billion years ago, to the present time. It is this possibility of actually directly observing the universe as it evolves through its history that has driven astronomers to build larger telescopes and more sensitive detectors, to put telescopes in space that enable us to detect more distant galaxies so that we may observe how the Earth, the Sun, and the Milky Way formed and evolved to their present states in a way that produced life. To what extent the *HST* might be able to surpass what the large ground-based telescopes had accomplished had been shown by the careful analysis of John Bahcall and his collaborators to be uncertain. It was this challenge that motivated the core group of Institute scientists interested in the distant universe at science coffee every day to push the envelope of galaxy evolution as far as they could, and they provided the impetus to direct the *HST* to this problem.

AAS | IOP Astronomy

Robert Williams

Chapter 6

Prelude to the Deep Field

One of the remarkable accomplishments of astrophysics in the past 30 years has been its use of advanced computing facilities to construct cosmological simulations. Researchers have taken the nearly perfectly uniform, homogeneous gas in the early universe with its minuscule fluctuations depicted by the cosmic background radiation shown in Figure 1.1 and have applied known physical laws and processes to that initial state to follow its evolution up to the present time, 13.7 billion years later. The computations are mammoth undertakings to be sure, as various groups have represented realistically the evolution of the hot, expanding gas producing the microwave background to determine how gravitational interactions among the components of gas, star formation, and radiative cooling interact to produce the large-scale structure of the universe.

Researchers have found that two effects dominate the evolution of the early universe: (1) the slightly over-dense regions succeed in gravitationally accreting neighboring gas so they grow in mass, which causes them to contract to higher density and become dominant. The large fraction of this gas does not interact with radiation and is called "dark matter". A small portion of this mass consists of atoms and as it heats up it radiates light. (2) A fraction of the gas avoids being gravitationally accreted due to its large distance from the nearest, more massive globules, and in the rapidly expanding universe that gas becomes cooler and progressively less dense. The resulting universe is therefore a mixture of different phases of matter: unseen dark matter; warm/hot gas of higher density consisting of atoms that forms gravitationally bound structures that can be observed—galaxies and clusters of galaxies on the largest of scales; and a very low density, cold "intergalactic" gas between the large galaxy structures that is essentially a void. After numerous galaxies have formed, some of the cold intergalactic gas is heated up by radiation from the young galaxies.

The cosmological computer models of the evolution of the universe from the earliest stages in which it has been observed, characterized by the microwave background radiation, prescribe how the cosmos evolves from its beginnings to the

doi:10.1088/978-0-7503-1756-6ch6

present time. These complex simulations require monumental effort and months of calculations on the most advanced supercomputers. A number of carefully developed simulations now exist, some of which are accessible on the web. They vary in specific details but virtually all of them show a similar pattern for the development of large structure in the universe: after some billions of years the expanding universe transforms itself from its very early caldron of hot, smooth uniform gas into a "cosmic web" of structure that consists of countless galaxies grouped in clusters. The endpoint of the simulations is, in fact, very similar to what astronomers do observe on the largest of scales in surveys taken with ground-based telescopes.

There is no better way to visualize how the universe evolves on the largest of scales than to view a cosmological simulation that, based on detailed calculations, presents its changes in basic structure over cosmic time in just a matter of minutes. One such recent simulation, executed by D. Rennehan at the University of Victoria, is shown as a video in Figure 6.1 (accessible at: http://iopscience.iop.org/book/978-0-7503-1756-6 for readers of the printed version of the book). The computational

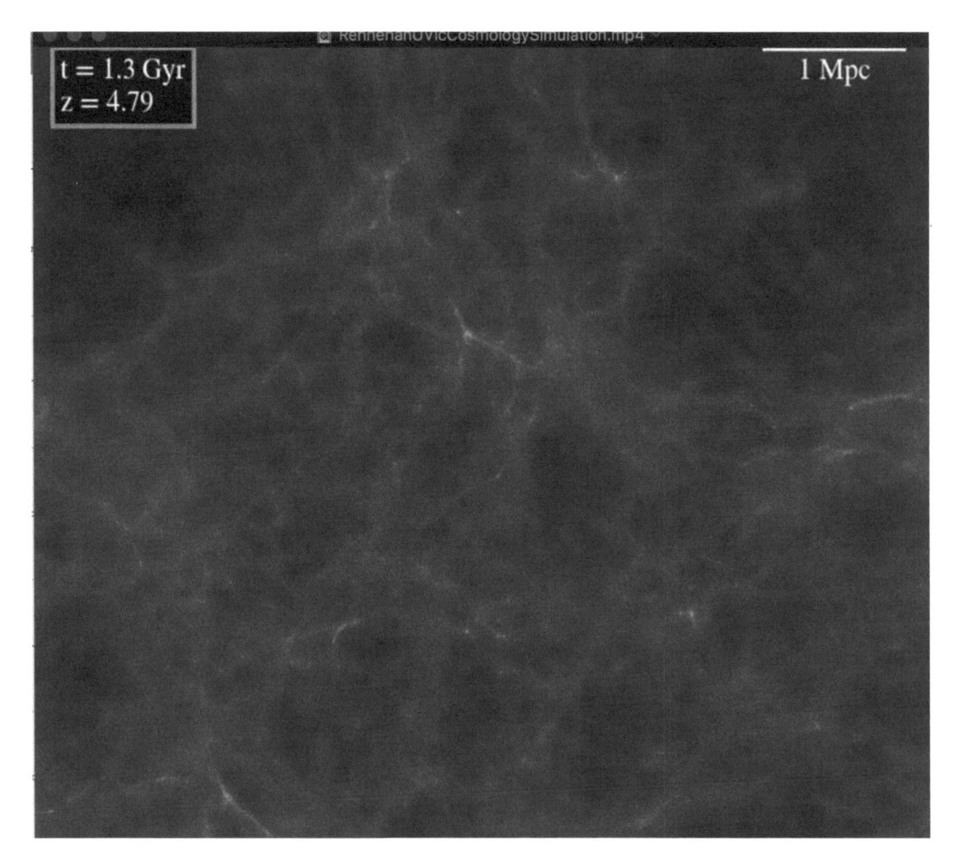

Figure 6.1. This cosmological simulation is a pictorial representation of the evolution of gas in the early universe from its initial homogeneous state through the formation of structure from interactions involving gravity and radiation. The bright nodes represent immense concentrations of matter out of which galaxies form, whereas the dark regions are largely devoid of matter [credit: D. Rennehan/Univ. Victoria]. Video available at http://iopscience.iop.org/book/978-0-7503-1756-6.

simulation begins at a time 50 million years after the Big Bang when cosmic gas is still quite homogeneous, and it progresses in rapidly increasing time steps to the present epoch, some 13 billion years later as structure assembles and becomes increasingly more prominent in the form of accreting filaments of gas. These filaments with their nodes of activity consist of both dark matter and the atomic, or "baryonic", gas that is shown in the video and which represent the radiating galaxies that are the prized beacons of the distant universe so diligently searched for by astronomers with large telescopes.

Although cosmological simulations were in their infancy in the early years of Hubble telescope's operation enough was understood about cosmic evolution that observations of distant galaxies in the early universe were one of the priorities for HST programs. Prior to the *HST's* 1990 launch, the community had gone through an exercise that defined three Key Projects to be carried out with the *HST* in its first year of operation, with commitments of substantial amounts of observing time on the telescope. Evaluated by a broadly based high-level review process, the Key Projects had been considered apart from the normal annual *HST* allocation procedure. One of the Key Projects, called the Medium Deep Survey (MDS) and headed by Richard Griffiths of Johns Hopkins University, was designed to use the WF/PC in a "parallel" mode to take images of whatever patch of sky was located next to wherever the primary *Hubble* instrument was pointed. It is possible to use two and even three *HST* instruments simultaneously, with each looking at adjacent regions of the sky. While one astronomer is using the Faint Object spectrograph to observe a star, the WF/PC can be intercepting another part of the *HST* beam to take an image of some other object that is near the star, for a totally different purpose.

The MDS was successful in producing numerous images of the sky that captured random objects such as stars and galaxies, most of which were not particularly noteworthy. The goal of the MDS was to ferret through the myriad of random images and identify those that showed faint, presumably very distant galaxies that might be of some interest for galaxy evolution and cosmology studies. This survey took place in the first 2–3 years of the *HST's* operation, during the time it was suffering from its spherical aberration. Nevertheless, the images were still superior to those that most ground-based telescope could produce. Figure 6.2 shows a montage of several of the galaxies that were imaged by the survey and subsequently confirmed from the redshifts of their spectra to be at distances of up to 4 billion light years, signifying that the galaxies are observed as they were 3–4 billion years ago. The objects in the figure show a morphology that is different from the appearances of nearby galaxies, and they quite likely give an indication of the shapes and sizes of galaxies at that time. The MDS results were provocative in demonstrating the ability of the *HST* to image distant galaxies, even with its spherical aberration.

An additional early *Hubble* program, devoted to imaging the more distant universe, consisted of a series of observations that had been approved for Alan Dressler and colleagues to study a cluster of galaxies that had been discovered by ground-based telescopes. At a distance of four billion light years, the galaxies appeared only as fuzzy blobs in large telescopes, and they were unusual in having

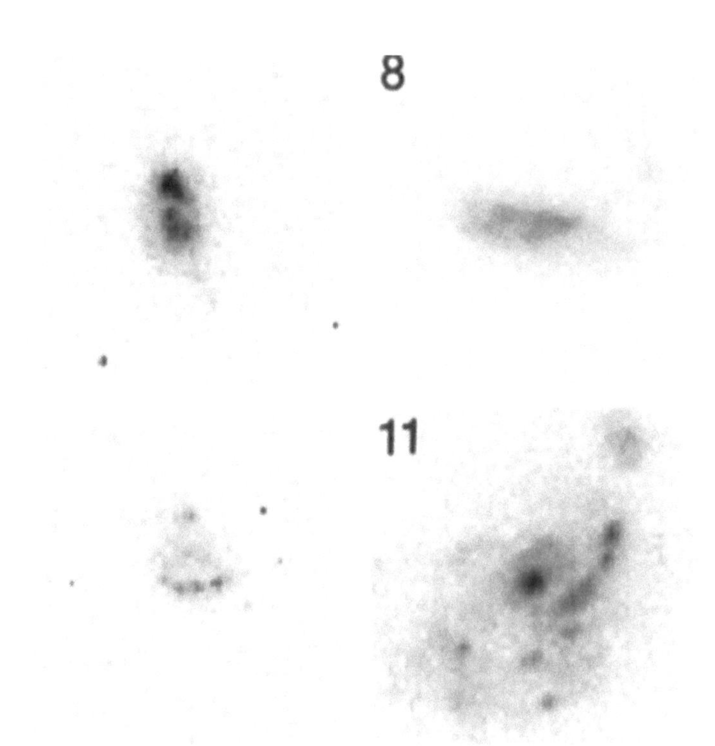

Figure 6.2. Random galaxies imaged by the Medium Deep Survey that subsequently were determined from their spectra to have distances of 3–4 billion light years. These galaxies are still in the process of evolution via mergers and accretion of intergalactic gas [credit: R. Griffiths].

quite blueish colors. Dressler and colleagues were interested in obtaining images of the cluster of galaxies to find out if the *HST* could reveal why those galaxies were bluer than galaxies at the present epoch of cosmic time. As with the MDS Key Project, the Dressler program to image the cluster, named from its coordinates in the sky as cluster 0939+4713, was initially carried out with the *Hubble's* aberrated optics using WF/PC. An image reconstruction method utilized by scientists, based on the concept of *maximum entropy*, used measurements of the *HST's* optical distortions to make corrections to the telescope's less-than-ideal images, such that some of the adverse effects of spherical aberration were corrected for. Like the MDS observations, the images were striking in clearly displaying galaxy shapes that did not show the better-defined symmetry possessed by galaxies near our Milky Way galaxy, in what astronomers call *the local neighborhood.* Rather, the galaxies of the cluster 0939+4713, observed as they were billions of years ago, are smaller and show what we would now consider anomalous features, e.g., multiple nuclei and asymmetric spiral arms. These observations were received with such enthusiasm by the extragalactic community that Dressler and colleagues were awarded time to repeat their imaging program of the 0939 cluster with the post-servicing mission corrected optics on the *HST*. A montage of their galaxies, shown in Figure 6.3, clearly demonstrates that significant differences in characteristics do exist between the CL0939 cluster galaxies and *local* galaxies.

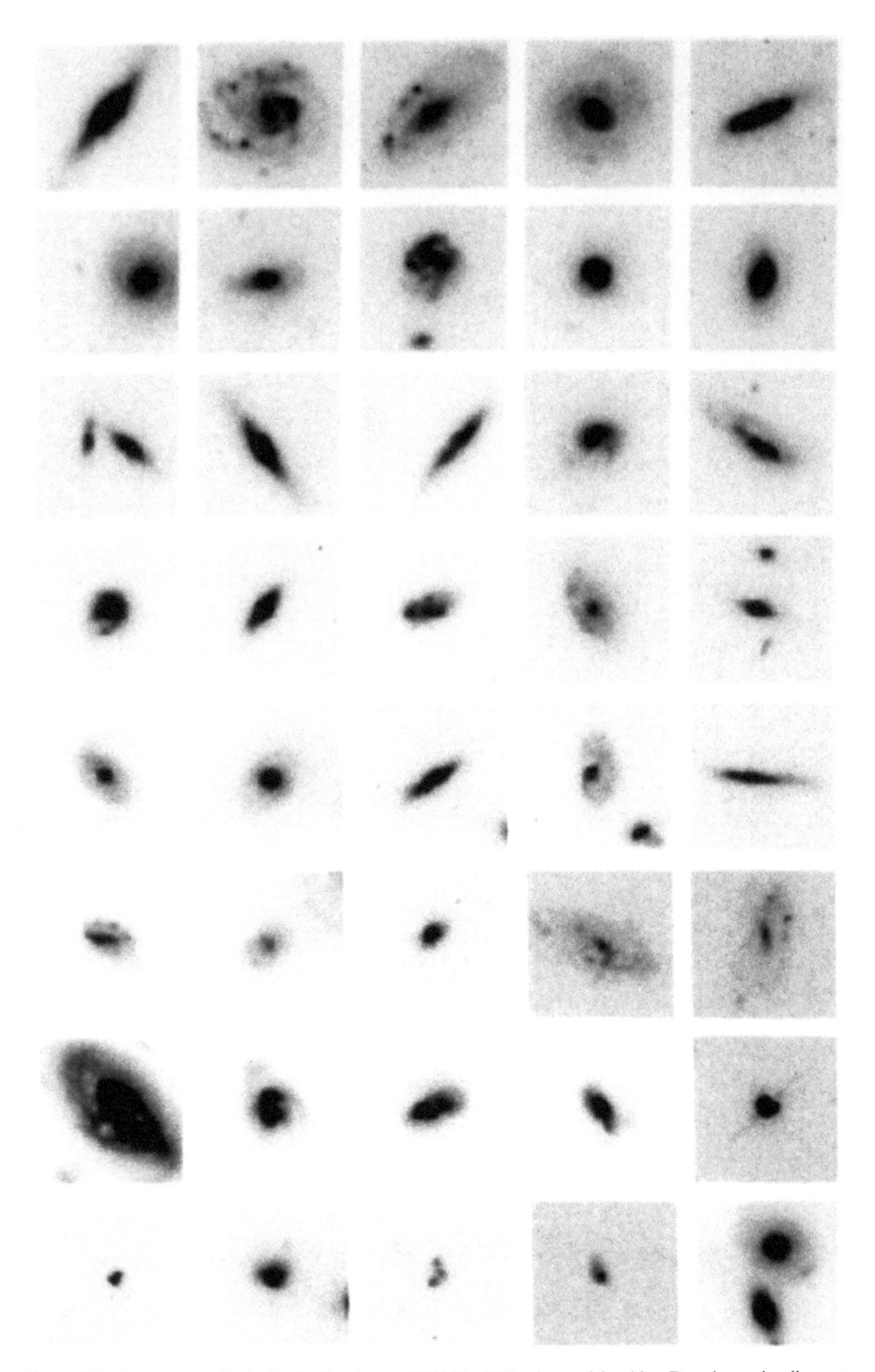

Figure 6.3. A montage of galaxies in the cluster CL0939+4713, observed by Alan Dressler and colleagues. They are smaller and more irregular than galaxies in the nearby, present epoch universe [credit: A. Dressler].

Nearby galaxies, which are bright and can be studied in detail, characterize galaxies in the present universe, 13.7 billion years after the Big Bang. Galaxies within a few tens of millions of light years' distance from the Milky Way are representative of the *local universe*. Although a distance of tens of millions of light years may not seem very local, it does represent looking back less than 1% of the time interval since the Big Bang. In the language of astronomers, this is the *local*, or present epoch. A montage of some typical nearby local galaxies that are relatively nearby in astronomical terms, i.e., on the order of a few million light years away, are shown in Figure 6.4. These galaxies are typical of the sizes and shapes of the larger, more massive galaxies that exist at the present time in the universe; they are likely to represent the evolved state of their predecessors, which may be bright enough to be seen through the largest telescopes at great distances. What stands out is that larger local galaxies have a regular structure. A bright nucleus is prominent and surrounded either by a smooth diffuse envelope of stars or a symmetric spiral structure. By contrast, the 0939+4713 galaxies have multiple nuclei and appear very clumpy, lacking the smoother appearance of the Milky Way and its neighboring galaxies.

The results of both the MDS Key Project and the *HST* imaging of cluster CL0939+4713 were significant in demonstrating that the renovated *Hubble* telescope should be able to provide even better observational evidence for basic changes in galaxy structure during the past four billion years and beyond. This author remembers well a conversation on a plane flight, seated next to Alan Dressler, during which he first showed me the galaxy images from his CL0939 observations; he urged that *HST* push forward with the imaging of even more distant galaxies.

A fundamental concept of the natural sciences is that everything evolves; nothing in the universe is truly static. The ineluctability of change comes not just from direct observations but from the fact that natural forces lead to the transfer of energy between all matter, and changes in energy produce a change in the state of an object. Physical laws prescribe how objects react in different conditions. By applying physical principles and our knowledge of the current state of an object, it is possible to project back to what the object was like in the past and extrapolate what it is likely to become in the future. This predictive process is limited by the fact that we are never sure if we know the true state of an object or that we understand all the relevant forces and sources of energy that influence it. Therefore, the Holy Grail of astronomy is to determine the evolution of major structures in the universe by observing them at different times so we have a directly observed understanding of how they change. It is the best way to determine what processes have been important in shaping our own galaxy, the Sun, and the Earth, and ultimately to understand what physical processes have produced the conditions that lead to the formation of different astronomical objects.

The age of the universe since the Big Bang is roughly 14 billion years, so if we are to observe astronomical bodies in different evolutionary phases, we must look at them back in time by billions of years. This requires observing objects at these huge distances across the universe. That's a tall order because even luminous galaxies are scarcely bright enough to be detected by the largest ground-based

Figure 6.4. Six galaxies that are relatively near the Milky Way and therefore typical of galaxies observed at the present epoch of the universe. They show an ordered symmetry, having spiral and elliptical shapes that galaxies in the early universe generally do not possess. Irregular galaxies, such as the one in the lower right panel, do have structures more typical of galaxies in the early universe, and it is believed that many of them have experienced phases of active evolution recently [credit: NASA/ESA/Z. Levay].

telescopes at these distances. At the time of the first servicing mission, when the *HST* was repaired to optimal performance, the aberrated images of the MDS Key Project and the CL0939+4713 program had already demonstrated that galaxies observed as they were four billion years ago did appear different from galaxies at the present time, in the local neighborhood near the Milky Way. The MDS and CL0939 cluster galaxies were faint, but not so faint that observing more distant galaxies was out of the question. Those programs were crucial demonstrations that the *Hubble,* with its restored optics, could well push back the distance to which galaxies could be observed.

Mark Dickinson was one of the dozen post doctoral associates at STScI in 1994, having arrived in Baltimore to take up his research position immediately after receiving his PhD at UC-Berkeley. His thesis research was an investigation of galaxies that emit strong radio waves. Extragalactic radio sources, as these objects are called, tend to be galaxies that have developed a very energetic environment that produces strong emission at long radio wavelengths. Dickinson's sample of galaxies came from a survey that radio telescopes at the University of Cambridge had undertaken in the 1970s to detect objects that were bright radio emitters. The Cambridge astronomers were collecting data to determine why certain galaxies emit such strong radio waves. They created several catalogues of numerous such *radio galaxies*: one particular galaxy in the surveys, 3C 324, was so named simply because it was source number 324 in the third Cambridge catalogue. Virtually all of the radio sources were either galaxies or quasars—extremely energetic sources of radiation now believed to be produced by gigantic black holes in the centers of their galaxies.

Dickinson had used large ground-based telescopes to characterize the structure of the radio sources from images and spectra. He focused his attention on the radio sources because the violent activity in most of them causes them to be very luminous (i.e., intrinsically very bright), so they were potentially visible out to very large distances. He was hoping to capture images at visible wavelengths using large telescopes on the ground, with the goal of detecting structure in galaxies that were more distant than had ever been imaged before.

During his thesis project, Mark had, in fact, discovered a cluster of galaxies that was at a much greater distance than any previously known group of galaxies. In 1993, observing with the 4 m telescope on Kitt Peak in Arizona, he had taken a spectrum of the brightest galaxy in the cluster of 3C 324, which appeared only as a tiny, very faint indistinct blob. On measuring the so-called redshift of the spectrum, he was amazed to find the galaxy cluster to be at a distance of nine billion light years—twice the distance of virtually any previously studied cluster of galaxies. The large distance of 3C 324 was a remarkable stroke of good fortune because it meant the galaxy was being observed at a time when the universe was only one-third its present age. Like examining an ancient fossil on Earth to piece together the evolution of life on our planet, Dickinson's immediate thought was to use the *Hubble* to take a series of long exposures of the cluster of galaxies around 3C 324 to see if the refurbished *HST* might be able to resolve the structure of the galaxy. What better way to piece together the evolution of structure in the universe than to see how physical laws shaped the processes that formed galaxies from the Big Bang to the time, 4.7 billion years later, when 3C 324 presented itself to our telescope?

As good fortune would have it, Dickinson's decision to image 3C 324 with the *HST* came right after the *Hubble's* first servicing mission had been scheduled for 1993 December. Anticipating the huge interest of astronomers to obtain observations on the *HST* that would surely come with successful correction of its spherical aberration, the Hubble Project at Goddard, ESA, and the STScI had just issued an international request for observing proposals that could be undertaken with the *HST* as soon as SM1 was successfully completed. Dickinson dutifully wrote up a careful observing proposal requesting 32 orbits of exposures for one pointing of the telescope at the central group of galaxies around its most luminous member, 3C 324. All of the exposures would be electronically stacked to create one image of these faint fuzzballs.

Mark's request for 32 orbits was far more time than anyone had ever been assigned for a single observation on *Hubble* telescope in the first three years of its spherically aberrated operation. It was a daring and gutsy proposal in which he, a recent PhD hardly known yet in the world of large telescope players, justified that it was necessary to push the *HST* to its limit in detecting such distant objects radiating barely above the galactic and cosmic backgrounds. Sometimes lightning does strike twice, and the Dickinson proposal was discussed and approved by the Hubble Telescope Allocation Committee in its Cycle 4 deliberations. There is an understandable tendency for review committees to favor proposals that are not so risky that they stand a good chance of not achieving their goals, but the committee that judged Mark's proposal favorably evaluated his careful assessment of the possibility of seeing structure in galaxies that inhabited the relatively early universe, and recommended the large number of 32 orbits for it.

The first science observations undertaken by the *HST* after SM1 had been completed were taken in late 1994 January. Many constraints dictate when specific targets in the sky can be observed with the *Hubble*. These include where the object is in the sky, the length of the exposure, how close the Sun and Moon are to the object, and whether the orientation of the telescope will allow it to observe nearby guide stars with the small star-tracker guide telescopes that keep the *HST* locked onto a target object. In the case of Dickinson's 3C 324 observations, the 32 orbits of images were carried out in 1994 May/June, shortly before the *HST* was to devote two solid weeks to the comet Shoemaker–Levy 9 collision with Jupiter, and they represented by far the most sensitive observations obtained to that time by the *HST*. Given that all the exposures taken during the 32 orbits were of the same target, using the same guide star, and through the same filter, the calibration of the individual images was expected to be straightforward.

It is crucial to reduce all sources of unwanted signal on the CCD detector to a minimum so that the faintest detail can be detected. As part of this process, it is important to take into account the small variations in sensitivity of each pixel in the detector. These are calibrated frequently as part of the normal operation of the telescope, via a process accomplished by imaging a uniform source of light across the detector that is produced by a special lamp. Spurious sources of (what appears to be) light on a CCD are often bogus because they are produced by cosmic rays striking the detector pixels, and these must be eliminated. All of this requires careful effort to clean the pixels, of which there are typically 1–10 million for every CCD chip. Algorithms exist that enable this process to be done for every exposure at a reasonable cost of effort and time. Once this process is completed, all the exposures are combined digitally to achieve the final image, which consists of only the actual radiation from the astronomical source.

During the early summer, when Dickinson received the data from his 3C 324 observations and was going through the detailed data reduction routines for all of his many images, he was in the process of moving from Berkeley to Baltimore to assume his new research position at STScI. Upon his arrival, he quickly fell in with other scientists whose primary research interests were other galaxies. In particular, Harry Ferguson, another of the bright postdoctoral researchers—who had been awarded one of the most highly prized research positions in astronomy, a coveted Hubble Fellowship—became one of his close collaborators. Ferguson shared not only similar research interests to Mark's, particularly the imaging of distant galaxies, but also an especially rigorous research style. Both were meticulous in their handling of data, not given to overstatement of the significance of any aspect of their own research, and they understood the numerous data reduction software packages that the Institute had created for *Hubble* observers. Both gregarious, Harry and Mark were regular attendees of the daily Institute science coffee and among the drivers of the energetic ambience there, where research results were discussed and ideas vetted. In a large Institute where details of the science operation of the *HST* necessarily occupied the majority of the time and effort of many of the scientists, these morning coffee klatches were a most stimulating part of the day.

Memories of the majority of scientific exchanges that took place during morning science coffees tend to fade away with time, overwritten by the most recent discussions that come up when one has a unique telescope at one's disposal. However, there is one moment that the few of us who experienced it can never forget. One morning in 1994 September, Mark Dickinson showed up at coffee in the library holding a glossy print in his hand. He announced to us, "Hey guys, I've just finished reducing the data from my 32 orbit *HST* program. Take a look at this WFPC2 image of the 3C 324 galaxy cluster." We gathered around as Mark held out the 8 × 10 inch black and white glossy (shown in Figure 6.5) that he had made of the combined image from his 32 orbit series of exposures of the cluster. For those of us familiar with the MDS and Dressler's CL0939 images, the picture was a knockout; it showed all sorts of detail for a number of the galaxies. Of greatest interest to us, it was clear that not a single galaxy looked anything like those that we were used to seeing in the local universe. They were all smaller, somewhat chaotic, and dysmorphic. We did not see any nicely symmetrical objects with smooth fuzzy elliptical shapes or spiral arms like those of our own Milky Way galaxy. Mark jokingly referred to the myriad of galaxies as "train wrecks" and "garbage galaxies."

Our first question for Mark was obvious: "What's the redshift of the cluster?" Redshift, like an object's magnitude (or brightness), is one of many astronomical terms that serve as proxies for basic, important information about the object. An object's redshift is the astronomer's way of asking its distance, i.e., its *lookback time*. Mark's response was, "It's at redshift 1.2." That indicated its distance to be nine billion light years, corresponding to the same impressively large lookback time. We were observing galaxies that clearly showed a diffuse, irregular structure at a time only 4.7 billion years after the Big Bang. The different forms of the 3C 324 cluster of

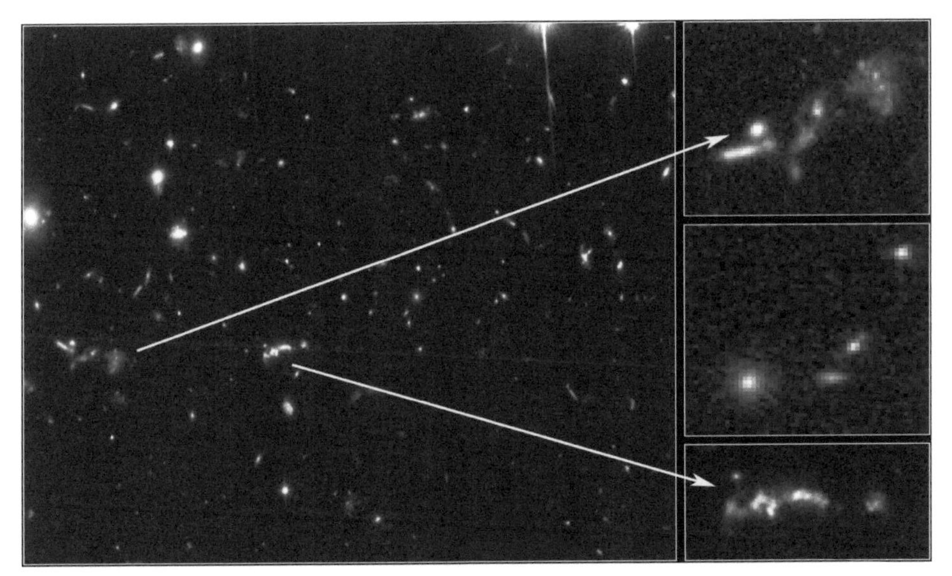

Figure 6.5. Mark Dickinson's 32 orbit image of distant galaxies associated with the central galaxy 3C 324, at a distance of nine billion light years. The right panels show several of the galaxies in detail. This landmark image demonstrated that the renovated *Hubble* could detect detail in galaxies at distances far greater than had been achieved before [credit: STScI/M. Dickinson].

galaxies, compared with modern-epoch galaxies, clearly demonstrated the consequences of galaxy evolution in the nine billion intervening years since the light we were observing had left those galaxies.

For those of us who had followed the decades-long effort to discover and image ever more distant galaxies, Mark's image was a stunning revelation. Even though my own personal research interests were more focused in another area at that time—stellar nova outbursts—it was one of the most transformative moments I can remember having experienced professionally. We were looking at galaxies that were significantly more distant than any that had ever before shown any structure—much more distant than the galaxies imaged by the Medium Deep Survey or those imaged by Dressler and colleagues in their cluster 0939+4713 at four billion light years' distance. For 3C 324, the cosmological effects that served to make galaxies appear increasingly faint with distance had been overcome by the *HST*. What we saw with our own eyes were unusual forms that made it clear that the *Hubble* could probe the distant, early universe better than anyone had foreseen. For the next few days following our initial look at the image of 3C 324, those of us at morning science coffee could hardly contain our amazement and excitement. We could talk of little else other than the fact that this image was a compelling argument that study of the distant universe must become a dominant theme for the *HST*: the question was how. The Hubble Deep Field Project was born at science coffee that week in September.

Robert Williams

Chapter 7

The HDF: Shot in the Dark

Ninety percent of *Hubble* observations are approved by a competitive annual peer review process in which scientists from around the world evaluate proposals that are submitted to STScI requesting observing time. Observing programs are normally proposed by collaborations of scientists who have similar research interests. Each year for the past quarter century, the amount of observing time on the *HST* requested by all proposals (of which there are typically more than 1000) submitted to the Institute has exceeded the time available by factors of more than five. Plain and simple, it is hard to get observing time on the *Hubble Space Telescope*. There is an understandable pressure among the peer review panels to approve a diverse slate of programs that are the most likely to result in discoveries and make the most significant possible advances in science. Because the total time requested for worthy proposals always exceeds the time available by such a large factor, one way to approve as many good research programs as possible is to award less time to each project than they have requested so that more proposals can get a slice of the pie. Observing teams are therefore often forced to focus on only their highest-priority goals, which detracts from and often even negates their ability to obtain the data needed to draw clear conclusions from their observations. One of the more common complaints heard in science after panel review results are announced is that the panel approved a proposal but did not recommend enough observing time or funding to accomplish what was proposed. In this situation, even if valid reasons exist for advocating a thematic agenda for the *HST*, such as emphasizing study of the distant and early universe, trying to influence the 130 independent scientists who constitute the annual peer review panels is not likely to be successful.

An alternative process exists for scheduling time on the *Hubble* that is under the purview of a single individual, and is therefore separate from the normal peer review process and the positive and negative aspects of its culture. The international standard by which scientific proposals requesting observing time and funding are judged is anonymous review by one's scientific peers. This is certainly

doi:10.1088/978-0-7503-1756-6ch7

the best procedure to follow, although at times it does introduce enough of its own biases to be counterproductive. When the number of panel reviewers is small, their collective judgment is more likely to reflect a few strongly advocated opinions that may or may not be constructive for the review. When the number of reviewers is large, as is commonly the case for review panels, compromises are frequently required for the group to arrive at a consensus, and this situation works against high-risk, high-return proposals. For precisely this reason, the existence of an alternative process for getting telescope time awarded by a scientist familiar with the telescope, preferably one who is not risk-averse, is not just a valuable option but a necessity for forefront science.

In the case of the *Hubble Space Telescope*, by its contract with NASA, 10% of the observing time on the *HST* may be scheduled by the STScI Director. His/her use of that "Director's Discretionary" (DD) time is reported to the Institute governing Council and to NASA, and is one of the criteria by which the Director's performance may be evaluated. For those of us who have been fortunate enough to have occupied the position of Director, assigning DD time is one of the great perks of being the director of a national center with a unique facility. The science that one can facilitate, especially for proposed observations that may be difficult to perform or are controversial, may be revolutionary. It may also be unproductive, but that must be accepted if one is to test new ideas. So it was that, within 24 hr of first seeing the image of 3C 324 presented by Mark Dickinson at morning science coffee, I was convinced that a sizeable portion of DD time available on the *HST* for the next year should be devoted to research focused on identifying and characterizing distant galaxies.

The idea of having a coordinated study of distant galaxies as a major research effort on the *Hubble* was hardly ground-breaking. However, it did represent only one topical area out of the many that were clamoring for time on the newly refurbished telescope, so there was likely to be pushback from the larger astronomical community. For several days, I batted around various ideas with our science coffee group about how best to devote a significant amount of DD time to the study of distant galaxies. Our collective discussions initially settled on two options. One option was to put most of the DD time at the disposal of the international community by announcing the opportunity for interested scientists to submit proposals for observations. I proposed that a special proposal review panel could be convened, one that would be separate and independent from the normal process of annual peer review of proposals. That panel would evaluate the proposals and recommend to me, as Director, those to be assigned *HST* observations. We were confident that such a plan would attract all the expert international galaxy researchers to participate, and excellent proposals would be forthcoming that would make successful use of significant amounts of *HST* observing time.

The alternative option that surfaced during our coffee discussions was for the Institute to form an internal team of staff scientists that could carry out a very large observational campaign on the telescope, focused on distant galaxies. An external panel would be named that would advise me as Director as to which important observations ought to be undertaken, and the observations would be carried out on behalf of the community by Institute scientists, who were most familiar with the *HST*. Both possibilities had attractive aspects, but the option of having an Institute group carry out the observational program did have an unattractive aspect—the *HST* exists for the benefit of the community, and having a large segment of *Hubble* time controlled by STScI scientists, i.e., insiders, might be resisted by the astronomical community, who understandably would want to be involved themselves in all aspects of projects to be done.

In truth, I initially felt that it would be best to give the discretionary time to peer-reviewed projects carried out by community teams consisting of experienced extragalactic astronomers. Our science coffee group that had been debating these alternatives was open to either option. As interested as they were in getting ahold of *HST* observing time for their own research, they realized the propriety of opening the DD time to everyone. They also knew that, as Institute scientists whose functional duties caused them to know more about the capabilities and nuances of the *Hubble* than anyone else, their expertise would likely result in them being invited to join community proposal teams.

Irrespective of how and for whom DD time was going to be parsed out, many questions arose that needed to be answered in order to determine the best way to go about making the most constructive use of the observing time. Should we limit the distant galaxy effort to a few projects, thereby allowing each one more observing time? Or would it be better to award fewer *HST* orbits to more programs? Which filters should be used? What types of galaxies or clusters of galaxies should be targeted? Should there even be specific targets, ones that might not typify normal galaxies, as opposed to studying undistinguished *blank* fields occupied by a random assortment of non-descript galaxies? Was it important to coordinate *HST* observations with observations on other telescopes sensitive to other wavelengths, e.g., infrared and radio telescopes, or X-ray satellite telescopes? Furthermore, because an image of galaxies is not useful for interpretation if the distances of the galaxies are not known, how could we go about getting spectra of the galaxies in order to measure their redshifts?

A related issue that arose at this time, one that was important to address, was the NASA/ESA policy regarding intellectual property rights. It decrees that teams carrying out observations on the *Hubble* are entitled to sole access to their dataset for a 12 month period before anyone else can have access to it. Would this policy, which protects a group's data from being seen and used by others, detract from our attempt to encourage a coordinated, broad-based effort in studying the early universe? It was a matter of some concern, although there was a catch that could provide a way in which data could be shared and analyzed by any interested persons.

Because the Director has sole control of DD time, s/he has the authority to approve only those observing programs where the proposing teams are willing to waive their proprietary data rights and make all their data immediately accessible to everyone. It was clear this would be a good way to foster a culture of open data, especially for a science that previously has had a history of keeping observational data proprietary under the exclusive control of the observer. Fortunately, the use of digital detectors on telescopes was already changing the paradigm of keeping one's data to oneself, because of the ease of storing and sharing electronic data as opposed to dealing with large, bulky photographic plates.

Some months went by as we presented the idea of assigning a large fraction of DD time to study distant galaxies to various groups, including the *HST* Project at Goddard and the STScI governing Council. Generally, the idea was met with support because of the positive response of the community to Dickinson's image of 3C324, which was made public in a press release in 1994 December. Theoretical studies of galaxy evolution and cosmology had been advancing impressively due to increasing computing power, and there was a strong desire to gather data on the early universe to compare with recent theoretical models that suggested how large structure formed and evolved. Dedicated observing programs on large ground-based telescopes had not been successful in cracking open the universe for direct observations beyond about four billion light years. There was a huge frontier that had been too difficult to pierce with existing telescopes. The *Hubble* had broken through that barrier in one 32 orbit exposure. It was clear that an extension of the *HST's* work in this area would be a boon for cosmology and galaxy formation.

Not surprisingly, there were objections raised in some quarters to the idea of assigning large amounts of *HST* time to a specific topic for which concrete results were not assured. Perhaps 3C 324 was unusually luminous and trying to observe other distant galaxies would not be so successful. Furthermore, the idea of having one individual, the STScI Director, make an important thematic decision for the *Hubble* went against the well-established tradition of having major decisions vetted via an extensive peer review process. I understood these concerns, but it surprised me that in a science culture that advocates diversity so strongly, there was such concern about the alternative to the normal peer review process of having a director assign DD observations for no more than 10% of the total observing time. I confess to believing very strongly in the wisdom of allowing the director of a facility to have some direct influence over the science that is carried out there. As good as the normal peer review process is, it does have shortcomings. There is every reason to have directors retain some authority over use of a facility as part of their responsibilities, and evaluate their performance upon how that authority is used. The fact that so many forefront telescopes and observatories have no official provision or process for DD time continues to baffle me. It is not in the best interests of science.

As the end of 1994 approached, *HST* observations were going well, but I was not certain of how best to move ahead with the idea of a concerted extragalactic program. I concluded that the wisdom of researchers having expertise in this area might be helpful in determining how to resolve the various scientific questions that our morning coffee group continued to debate. Therefore, in 1995 January, I decided to convene an external advisory panel of knowledgeable astronomers whose careers had been devoted to the study of galaxies. I asked them to come to the Institute for one day to give the Institute their advice on the best use of *HST* time to gather data on the early universe. Each panelist was invited to speak for 15 minutes, presenting their thoughts on the subject, after which there would be a full discussion of the best way to proceed in the allocation of the hundreds of *Hubble* orbits that would be offered for distant galaxy studies.

Before the panel meeting took place, our coffee group had already tossed around many of the relevant topics about how the *HST* could really make a mark in addressing the evolution of galaxies. We had formed our own opinions on the pros and cons of which filters to use, exposure times, possible target fields, collaboration with radio and infrared facilities, and how we might get spectra of galaxies imaged with the *Hubble*. So, we all had our own ideas of what would be good ways to proceed, and we were therefore very interested to hear what recommendations would emerge from the external advisory panel.

On Friday, March 31, the Advisory Panel met in Baltimore at the Institute, and the primary issue that received the most attention and energy that day concerned the question of observing targeted versus non-targeted areas of the sky—a topic on which strong opinions were voiced. Targeted areas, i.e., areas where there were already known to be moderately distant galaxies in clusters, at least guaranteed that *HST* images would contain objects to be imaged. On the other hand, such areas of high galaxy density were likely, by their very nature, to not be truly representative of much of the universe. Pointing the *Hubble* toward a random area of sky not known for any particular features or objects would be more likely to represent "normalcy." On the other hand, one could expend quite a lot of time imaging such a random area and capture very few objects.

According to the notes I took at the meeting, the panelists who voiced the strongest opinions on the subject of targeted versus blank field included Sandra Faber, Richard Ellis, Len Cowie, Alan Dressler, Simon Lilly, and Garth Illingworth. They offered rather divergent opinions on the subject and justified their points of view by drawing upon their own experience and various research results. After considerable discussion on the topic, no consensus among the panel emerged; each option—targeted versus blank field—was either advocated or objected to by roughly equal numbers of the panelists.

The other issues taken up by the advisory group involved the question of proprietary data rights and archiving the data, how many fields to observe, and which wavelength bands would be most useful. Again, there was an interesting discussion, but little consensus emerged from the panel of experts. As eminent galaxy researcher Richard Ellis reflected in an article he wrote two years after the meeting, "Those present will remember a rather rambling discussion with...much disagreement on details. We hardly prescribed HDF at that meeting." Indeed, the most significant outcome of the advisory committee meeting was not a series of recommendations to the Director so much as an informal and useful interaction with those staff of the Institute who were interested in galaxy research. At the end of the day, a number of the advisory panel members admitted to a certain frustration and disappointment that they had not come to any agreement on a strategy that would be useful in determining how best to use a large amount of *HST* time to study galaxies. I thanked the panel for their time and perspective, then we adjourned and went our separate ways.

The fact that the expert committee had not found concrete agreement on how best to prescribe important components of a program to probe the distant universe was a surprise and disappointment for me. I had expected convergence on many of the topics. In reflecting on this the weekend following the Advisory Committee meeting, I realized that I had failed, as convener of the panel, to pose the right questions to them. Worse yet, I had not effectively led the afternoon discussion. It was my primary responsibility, and for no particular reason other than not maintaining adequate focus, I had fouled up the panel meeting. I was very frustrated, especially with my performance. Still, I remained convinced that a viable program must be crafted using the several hundred *HST* DD time orbits available to carry out what our group believed should produce a ground-breaking glimpse of the early universe.

Over the weekend, away from professional distractions, I returned to thinking about the idea that our morning coffee group had kicked around some months previously. Why not take the matter into our own hands on behalf of the entire astronomical world and create a large program among ourselves to image the sky as deeply as possible with the *HST* and make that data instantly available to everyone? We could provide both raw and reduced data, i.e., data that had been calibrated and had instrument signatures removed by our expert Institute staff. The reduced data, which normally take considerable time and effort to produce, would enable the entire international astronomical community to begin analysis immediately without going through all the labor of doing what our Institute staff could do better and much more quickly than anyone. The idea had appeal to all of us. We wanted to be involved in pushing the boundaries of what the *Hubble* could do, but we also realized that we could not, as a national center created to serve the research community's needs, do anything that would favor our research interests at the expense of our colleagues worldwide. Whatever we did had to be done in a way that put all astronomers, whether *HST* observers or not, on an equal footing. In my mind, this was an efficient way to proceed that was as good as any alternative.

Early the next week, I informed our morning science coffee group that the decision was made: we should proceed in formulating a deep imaging observation and I would assign the remaining DD time to our group to carry it out on behalf of the entire community. Immediately, the group began strategizing on ways to formulate a concrete plan to use most of the DD time still available that year. All of the original group that had been part of discussions for the previous six months were eager to be part of creating a plan for coordinated observations on the *HST*, one we hoped would yield results even more exciting than the image of 3C 324.

We decided to move in two parallel directions. As Director, I would deal with the administrative and political issues that would need to be addressed because the Institute was creating a large program that was entirely internal to the Institute, albeit done on behalf of the community, over which we had complete control with no direct external scientists' participation. The other members of our core HDF Team, as we called it, Drs. Ferguson, Dickinson, Fruchter, Giavalisco, and Postman, would come up with different ideas that we would all consider as possible observations for roughly 200 orbits of DD time. We would move the bulk of our discussions away from the informal morning science coffees and begin meeting every few days in the Director's office to chart our path forward.

The program that eventually emerged as the Hubble Deep Field came about in rather short order from our discussions in early April. The key to how it was defined hinged on one central question: should we focus on observations of suspected groups of galaxies that might be at large distances? The first option was to continue with Mark Dickinson's practice, which had produced the image of 3C 324, of going after known objects that are likely to be strong radio sources because they are typically evidence for very energetic activity and therefore galaxies that are detectable at very great distances. The alternative, because radio galaxies are not representative of most galaxies, would be to try instead to observe typical, run-of-the-mill galaxies even though they might be less likely to be bright enough to be seen. The latter would give a better understanding of the nature of the universe at the epochs the galaxies are observed. It was this issue that had stirred the most intense discussion among the advisory panel members, and it held the most risk. Our team debated this issue at length, finally deciding that it would be important to observe a "typical" region of the universe, even though it made such an observation rather chancy, because of the possibility that there might be few, if any, distant galaxies appearing in an image of a "blank" field of sky. Such an image would truly be a "shot in the dark," but that is exactly what should be the proper image for students of galaxy evolution and cosmology to interpret.

The decision to select a blank field(s) for deep imaging, with its attendant risk of coming up largely empty-handed, would almost certainly not have been possible had this project required approval by a peer review panel. The temptation of making sure that a deep image requiring substantial amounts of time on the *HST* did contain sufficient numbers of distant galaxies would likely have pressured any normal panel to constrain such an observation to a field where it was at least assured that a distant group of galaxies was known to occupy the field. The fact that we did bypass the normal peer review process to schedule risky observations on the *Hubble* was essential to the Deep Field Project moving forward.

Following the decision to image a blank, undistinguished, previously unstudied spot of sky, other necessary decisions about the Hubble Deep Field, or HDF program as we were now calling it, fell into place. The next big decisions to be addressed were: how many blank fields should we attempt to image? What wavelength bands should we use? One can isolate different wavelengths, or colors, in the visible spectrum by using filters, and they serve the useful purpose of conveying information about the nature of the objects being photographed. For example, hot stars tend to radiate more light in the short (blue) wavelength region of the spectrum, whereas cooler stars radiate much more light at longer (red) wavelengths. Therefore, taking images of the same spot of sky with both a blue and a red filter to isolate these wavelengths would give an indication of the fraction of stars that are hot versus the fraction that are cool. This might seem trivial, but there are fundamental implications that follow from the hot versus cool star population of galaxies. Hot stars have short lifetimes, and therefore when they are observed, they must have formed recently. Cool stars have much longer lifetimes, and are therefore more likely to have already been in existence for billions of years. Hot stars give us a measurement of the rate of formation of gas clouds into stars, and this information is an important indicator of the history of the galaxy those stars belong to. We judged this argument to be strongly supportive of having Deep Field images taken in multiple wavelength bands.

Astronomical imaging in different wavelength bands does provide key information, but it comes at a price. Photons of light from wavelengths other than those passing through the filter band are not received, which causes objects in the image to appear fainter than they would if all wavelengths that they radiate could be detected by the instrument. Just how much weaker the objects would appear was central to our debate as to which and how many filters to use for the HDF. The more filters used, the more information we would obtain for the objects; however, the images would be weaker. After extensive discussion, we decided to take images in four wavelength bands spanning the full spectral range that the *HST* could detect. We would have images in ultraviolet (UV), which is not possible from beneath the Earth's atmosphere, and in blue, green, and red wavelength bands. One way to compensate for the weaker images that would result in each band would be to eventually digitally add together, or stack, the images in all the different wavelength bands, thereby producing a deep image with fainter detection limits—but also making possible the impressive impact of a full color image of the field.

The decision to image a deep field in multiple pass bands rather than accepting photons from all wavelengths meant that longer exposure times would be required to reach the faintest detection limits in brightness. Thus, our decision to concentrate on imaging blank fields placed greater importance on achieving the deepest possible detection limits in our exposures because of the uncertainty of what, if anything, might be detected beyond the few bright nearby galaxies already known to occupy the field. Calculations done by the group demonstrated that 25–30 orbits of *HST* time would be required to image a dark field down to the limits where background and detector noise began to dominate exposures taken with the WFPC2 instrument that we would be using. Because only 150–200 orbits of DD time were available for our deep field program and we wanted to obtain images in four different wavelength bands, obtaining the requisite exposures of a single area of sky would use up most of the orbits available. This fact imposed one of the other important constraints on our program: our observations to detect distant galaxies should be focused on just one field. We came to this key conclusion in late April, but there were still major decisions that needed to be made before observations could begin.

Chapter 8

Planning the Observations

The *HST* circles the Earth in a low orbit every 96 min. At its height, 600 km above the surface of the Earth, one full hemisphere of the sky is obscured by the planet at all times. The other hemisphere is the accessible dark sky. Even when the telescope is in full sunlight directly above the sunlit Earth, the sky overhead is quite dark because of the absence of atmosphere, so stars, planets, and galaxies can still be seen. Therefore, the *HST* makes observations at all times, 24 hr a day, except for those parts of orbits when the telescope passes through the South Atlantic Anomaly. That, however, does not mean that the *Hubble* can observe specific objects for the full duration of its orbital period. This is because, wherever the telescope is pointed in the sky (with the Earth below it), 48 min later the telescope will be on the opposite side of the Earth, so the planet will then be blocking the view in that direction. For this reason, most *HST* observations consist of series of exposures that are repeated when the telescope is in a part of its orbit where the object is visible, i.e., not blocked by the Earth. Those exposures are then digitally added together, with each individual exposure lasting for roughly half the telescope's orbital period, i.e., around 48 min, when the line of sight to the object is not blocked by the Earth.

For what fraction of the *HST's* orbit an object is blocked depends on the direction of the object relative to the plane of the telescope's orbit. An object in the plane is eclipsed by the Earth for a full 48 min. However, as the direction to an object moves away from the plane and toward the pole of the orbit, objects can be seen for a greater fraction of time. For objects that are in the same direction as one of the poles of the orbit, where the telescope is pointed along the limb of the Earth and grazing the top of the atmosphere, those objects are not occulted at any time as the *HST* orbits the Earth. There are two small cones, pointed in the respective directions of the north and south poles of the telescope orbit, having a narrow opening angle of only 3°; any objects situated within these cones can be observed continuously without obscuration by the Earth. These cones are called the *HST's Continuous Viewing Zones* (CVZs), and they present an opportunity for uninterrupted

doi:10.1088/978-0-7503-1756-6ch8

observations of stars and galaxies that are located inside the cones. Thus, objects in the CVZs can be observed significantly longer than those in other locations outside the cones, allowing more data to be acquired.

The value of studying astronomical objects inside the CVZs, where *HST* observations have the highest possible duty cycle, was one of the constant themes of my research colleague at the Institute, Ron Gilliland. Ron was persistent in advocating the value of the CVZ in making *HST* observations in order to maximize open exposure time for objects. This situation suited the Deep Field Project perfectly. We had decided to image a non-descript field of sky, trying to detect the faintest possible galaxies in it. Why not select a field inside one of the CVZs where we could expose for the longest possible time without suffering interruptions because of Earth occultation? It was an interesting idea to pursue, and the task of coming up with a good field location for the HDF fell to Marc Postman and myself, as leads.

In many respects, the decision on which exact spot in the sky to image for the Deep Field was one of the most difficult because we had constrained it by our decision to go with just one field and to locate it in one of the CVZs. It also need be mentioned that the *Hubble's* narrow CVZ cone actually sweeps around the sky slowly because the orbit of the *HST* precesses, i.e., it slowly gyrates around due to the fact that the Earth is not exactly spherical. The Earth's gravitational field produces a slight torque on the orbit of the *HST* that causes its orbital gyration, or precession, like a spinning top around the Earth's north pole every two months. The *HST* was launched in a due eastward direction from Cape Canaveral, whose latitude is 28°, and this produces a tilt in its orbit such that the orbital pole sweeps around in a circle above latitude 90 – 28 = 62°, corresponding in astronomical terms to a declination of 62°. The task Marc and I faced was to recommend to the HDF team a suitable patch of sky in this limited region.

The best camera on the *HST* for imaging a deep field was the Wide Field/Planetary Camera 2 (WFPC2), which had been installed on the first servicing mission to correct spherical aberration. Developed at NASA's Jet Propulsion Lab by an accomplished team headed by John Trauger, it contained a mosaic of four CCDs that provided a field of view of 2.4 arcmin across. That area of sky corresponds to only 1/25th of a degree, about the size of the period at the end of this sentence when viewed at arm's length. With our marching orders in place, we went over the requirements that a field needed to satisfy in order to be considered optimal for the Deep Field.

The HDF needed to be far away from any bright stars whose scattered light could dominate and obscure faint objects in the field. Similarly, because it would be important for observations that would be made at other wavelengths, the field should be far from any bright radio and X-ray sources that would make detections of the faintest galaxies difficult for those wavelength regions. For interpretation of the data from the HDF, it would be important to obtain the spectra of as many galaxies as possible that appeared in the image, and because the Keck 10 m telescope in Hawaii was by far the largest optical telescope in the world at that time, the HDF would have to be in the northern CVZ; the southern CVZ is not visible from

Hawaii's northern latitude. Crucially, the CVZ field for the HDF needed to be in a region of sky where there was as little background radiation as possible from clouds of gas and dust in the Milky Way, where observing through even the faintest diffuse interstellar cloud could produce a slight glow on the CCD that would render faint objects undetectable in the cloud background.

Fortunately, catalogs were available from previous telescopic surveys of different types of astronomical objects throughout the sky; these would tell us the exact location of any objects that would compromise the quality of the HDF if they were too near. These included radio sources, gas and dust complexes, and X-ray sources, so the task was straightforward: find a small patch of sky not near anything of any significance within the Milky Way and the sweeping cone of the northern CVZ. We wanted a very isolated dark spot—one as *blank*, i.e., nondescript, as possible—that would give us a bore hole through to the most distant universe. Marc attended to this diligently, and by the end of April, only four weeks after the advisory panel meeting, he sent me nine potential fields that he had identified as satisfying all our requisites. In his email, the candidate fields were labeled numerically 1–8, except for the one site that he considered best, which he designated "A." That evening, I grabbed my personal copy of the Tirion Sky Atlas and inked in the positions of the nine candidate HDF fields to see where they were in the sky. The locations of the candidate HDF fields are shown reproduced in Figure 8.1, all of them placed either in or very near the northern CVZ. The HDF

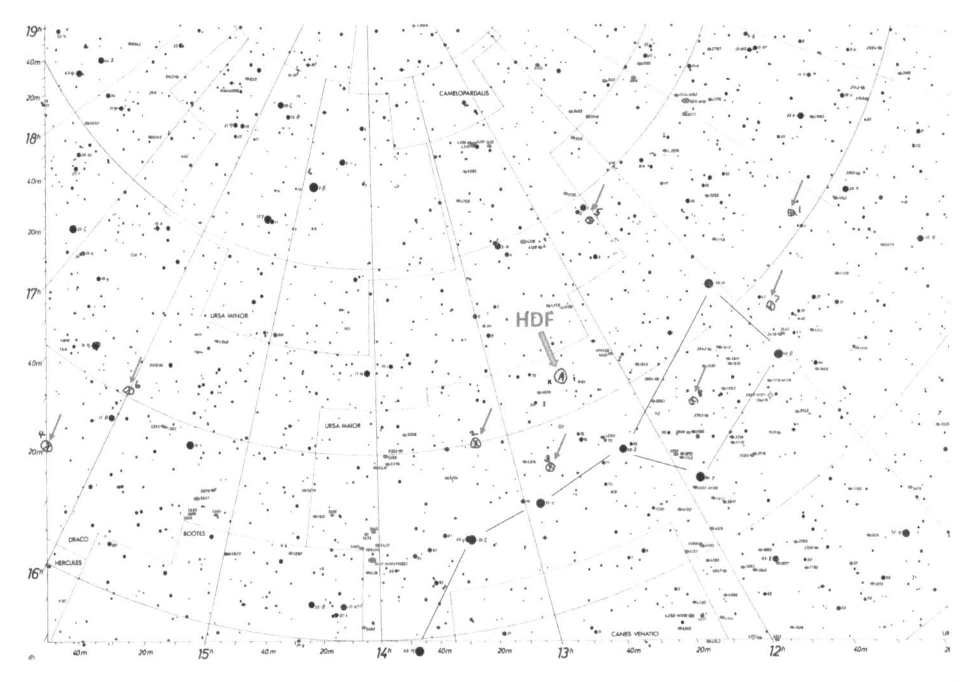

Figure 8.1. The locations of the nine HDF candidate fields in the northern CVZ are shown marked on the author's personal sky atlas in blue ink with the symbol ⊗. The most favored field, which was ultimately selected for the HDF, is just north of the Big Dipper and denoted with an "A" [credit: Sky Pub Corp/R. Williams]. A larger version of this figure is available at http://iopscience.iop.org/book/978-0-7503-1756-6.

Team reviewed the characteristics of the nine fields and agreed that site A appeared to be the best in terms of having the lowest background radiation and absence of nearby objects in our own Milky Way galaxy.

A final condition needed to be satisfied before the exact position of the HDF field could be specified; it had to be near an official *Hubble* guide star. When any telescope takes an exposure of an object that is fixed in position, such as a star or galaxy, the telescope pointing must remain absolutely fixed on it. Any change in the pointing of the telescope during the exposure will cause a blurry image. The *HST* accomplishes "star lock" with three independent optical detector systems called Fine Guidance Sensors that use small telescopes to image guide stars and lock onto them. The locking process is accomplished by the detectors sensing the individual pixels that are receiving the light from the guide star. If the telescope moves the slightest bit from its pointing position, it will cause the guide star image to move to a different pixel on the detector, which then creates an electronic error signal. The error signal triggers software that immediately brings the telescope back to its original position. This positioning loop occurs constantly on timescales of tenths of a second for all *HST* exposures and is the key to the *Hubble* maintaining its sharp images.

The requirements for *Hubble* guide stars are constrained. A guide star must have a certain brightness level, not be a binary star, and not be too close to another star in the sky. Binary stars move in position as the two stars circle each other, which negates their value of having a fixed position, and stars too close to each other can complicate the error signal loop. A great deal of effort has gone into identifying *HST* guide stars because many of them are necessary, distributed around the entire sky. No viable *Hubble* exposure can be taken unless there are several guide stars available that the Fine Guidance Sensors can detect. Guide stars will also be necessary in the future for virtually all space telescopes, hence the catalogue of acceptable stars still continues to be augmented and now numbers hundreds of thousands of stars in the Milky Way.

Once we had decided upon site A as the field of choice within the CVZ, we checked for guide stars that would be appropriate for our series of exposures over many days. There were a number available, depending on the precise position we selected for the HDF field. Because it was important to test the suitability of those guide stars before committing to the long series of HDF exposures, we made the decision to take a trial two-orbit observation of the specific field we had selected. The two-orbit observation would validate the exact positioning for *HST* and it would also serve the important purpose of detecting the brightest galaxies in the HDF. Because spectroscopy of the brightest galaxies in the HDF was essential for the analysis of the galaxies in the image, the two-orbit observation would be given to Keck Observatory observers, who had already committed to obtain spectra of the brighter galaxies on the 10 m telescope months before the full HDF exposures were to be obtained.

As circumstance would have it, the mid-June two-orbit preliminary observation of the HDF failed. The *HST* could not obtain a proper lock using the existing guide stars for the exact pointing we had specified in the small site A field. One of the guide stars had been assigned an incorrect position and there was a question as to whether that star provided an adequately clear image to create a valid error signal. Institute staff thought that the problem was fixable, but our HDF team did not want to take any chances. Other guide stars were available within the slightly larger prime site A field and they could be accessed by shifting our original HDF position by a very small amount. This we did, repeating the preliminary observation some weeks later, which turned out to be successful. The exact HDF field had now been specified and the two-orbit image was provided to interested observers who had access to the Keck 10 m telescope and were willing to obtain spectra of the HDF galaxies visible in that image.

Once the precise position and telescope orientation for the deep field were specified, most of the details of the series of observations that would constitute the HDF had been worked out, so the observations were ready to be scheduled. There were two remaining parameters to be decided—the total exposure (or observing) time and the dates of the observations We wanted to expose as long as was necessary to resolve the detailed structure of the galaxies. There does come a point in time for any exposure where noise from background sources and the instrument electronics begin to diminish the contrast between galaxies and unwanted noise. The team had calculated that it would take roughly 150 total orbits to reach this limit with observations in the four wavelength bands we desired. This number of orbits was fortuitous because the precession of the *Hubble's* orbit causes the narrow CVZ cone to sweep slowly around the sky such that any one point in the sky remains within that cone, i.e., is observable within the CVZ, for a maximum of only 10 days, or 150 continuous orbits. During that time, the CVZ cone continues its slow precession around the Earth's polar axis, returning back to the same point 56 days later. Thus, one can make observations of objects in the CVZ for 150 continuous orbits at most. This was roughly the number of DD time orbits that were available to use during the 1995/96 season, so that set the number of orbits we had available for the HDF.

As for the dates of the HDF observations, normal *HST* observations are scheduled automatically by software that takes into consideration factors that maximize the efficient use of the telescope. For example, the *HST* uses flywheels to change its orientation, not small thrusters whose expulsion of combustible gases would contaminate the telescope and instruments. The movement of the telescope is slow; it changes its orientation at the same rate as the minute hand of a clock. It is therefore more efficient to schedule observations of targets that are near each other in the sky, minimizing the down time that occurs while the telescope changes its pointing direction between observations. Complex software handles the scheduling problem by taking the large number of observations approved annually by the peer review process and sorting them all out, so the maximum amount of time can be spent with the instruments in operation, i.e., taking exposures. The one exception to this process occurs when observations are made in the CVZ. Because the CVZ is really a slowly moving, narrow funnel as it precesses around the Earth's polar axis, observations in the CVZ must be made when it passes over the object to be observed in its 56 day cycle. In the case of the HDF in late 1995, the only continuous interval when our field could be observed was December 18–28.

When word began to get out to the astronomical community in late April/May that the STScI Director had decided to undertake a very large observation whose success was doubtful and would use up most of the yearly allotment of the *Hubble's* DD time, the news did not go unnoticed. Because the Project had been formulated by and under the control of Institute scientists, not astronomers in the broader external scientific community, it did touch on certain sensitivities that had been around since national centers in astronomy were first created in the 1950s. The dichotomy was relevant to the HDF because the STScI, whose mission is to facilitate science in collaboration with external scientists and institutions, was creating a program on the *Hubble* via a team that did not include external scientists. More than anything, this was done to facilitate clear decisions and rapid action. External advice was requested and encouraged, but the crafting of the HDF program was done entirely internally at the Institute.

To ensure that the HDF Project had the imprimatur of NASA and the Institute-managing oversight Council that reported to AURA, I presented the plan for the HDF to a number of groups in the late spring/summer of '95. The *HST* Project at NASA/Goddard was very positive about the HDF when they were informed. *HST* Project Scientist Dave Leckrone, a strong advocate for *HST* science throughout its history and an effective supporter of the Project, was supportive of the idea that the *Hubble* could break through a historic barrier if it demonstrated its ability to detect distant high-redshift galaxies. The Space Telescope Institute Council, consisting of prominent scientists external to the Institute, was created by AURA to approve policy for the Institute and oversee its governance of the *Hubble* science program. When the HDF Project was presented to them in their June meeting at the Institute, a lively discussion ensued. Council Chair Ed Turner, astrophysics professor at Princeton, was quite enthusiastic about *HST* committing to an extensive program to study galaxies—this happened to be his active research field. Several other Council members had some concerns that the HDF would be perceived as a program crafted by the Institute that would reflect Institute interests rather than those of the community. However, when informed of the results of the advisory panel meeting, those concerns were mostly (but not entirely) allayed.

The most significant comment at the Council meeting came from Lyman Spitzer, also of Princeton and in many regards the father of the *Hubble Space Telescope*. Lyman, ever diplomatic, asked if the Institute was sure that it really wanted to do such a project with its risk of failure and the consequence of having the public turn further against a telescope that had already had serious optical problems, which had been fixed at great taxpayer expense. Prof. Spitzer was clearly concerned by what the failure to detect many galaxies might mean for the long-term future of the *HST*. When I answered him that Institute scientists felt it compelling that the *HST's* unique imaging capabilities be pushed to the limit, he merely nodded his head, smiled, and looked down. His concern was evident. Lyman approached me during the next coffee break and admitted his worry that the HDF Project I had described might not be the right program at that time to devote so much telescope time to, when a positive outcome was not assured.

AURA President Goetz Oertel, to whom the Council reported and who was my direct boss, attended meetings of the Council. After that meeting, he approached me. Acknowledging the significance of Lyman's concerns, he privately offered his unconditional backing for the HDF, even though we both knew it could end up as a campaign that might well elicit sighs of "Nice try, but..." Always mindful of Goetz's keen sense of the astronomical landscape and his unequivocal support, I assured him that, if the HDF turned out to be an embarrassment for the Institute or the *HST*—i.e., a result of poor judgment and execution by the Director—I would fall on my sword and resign the STScI Directorship. Goetz scoffed at the idea, but I was serious. The disastrous public and political reaction to the *HST's* spherical aberration taught us that undertaking a very risky venture at this point would require a legitimate focal point for any outcry over HDF failure that could at least preserve *HST* and space astronomy.

Generally, as news of our plan for the HDF diffused through the research community, most of the comments were supportive of what we were trying to do on behalf of extragalactic astronomy. There was one notable exception, however, and that came from John Bahcall. Three days after the June Institute Council meeting had adjourned I received a call from John. He wished to come down to the Institute from Princeton the next day, and wanted to know if we could have lunch together. John said the primary reason for his visit was to show me some interesting preliminary results of observations he had made with the *HST*. He also mentioned that he had heard about our plans to do the HDF and he wanted to share his perspective about it with me. I knew of John's reservations about the *HST's* ability to resolve distant galaxies, based on his analysis in the 1990 *Science* paper. I also knew that John and Lyman Spitzer were very close, and Lyman had just returned to Princeton from the Council meeting where he had already expressed his concern about our chances for success. The two had no doubt talked, and given John's impeccable research credentials and profound investment in the *HST*, he was important to listen to.

John's visit to the STScI was partly motivated by the research project he was working on with a team that had used the *Hubble* to image quasars whose distances were between 3–6 billion light years away. Refining the observations his team had made to detect faint light surrounding bright quasars involved complicated procedures, and STScI staff understood the characteristics of the *HST* detectors. John wanted to meet with Institute *HST* instrument specialists to assure that he and his team were removing instrument signatures from their data properly in their attempt to detect extremely faint galaxies associated with the quasars. John's diligence to details was ultimately vindicated. His group achieved impressive results that were subsequently published, as shown in the Figure 8.2 mosaic of quasars his group observed. They demonstrated that quasars are not isolated objects separate from galaxies, but are energetic regions at the centers of galaxies that can best be understood in terms of the complex environment where massive black holes are now known to exist.

My lunch with John did take place the day after he had called, and our conversation was devoted to some matters in science politics, as well as the data

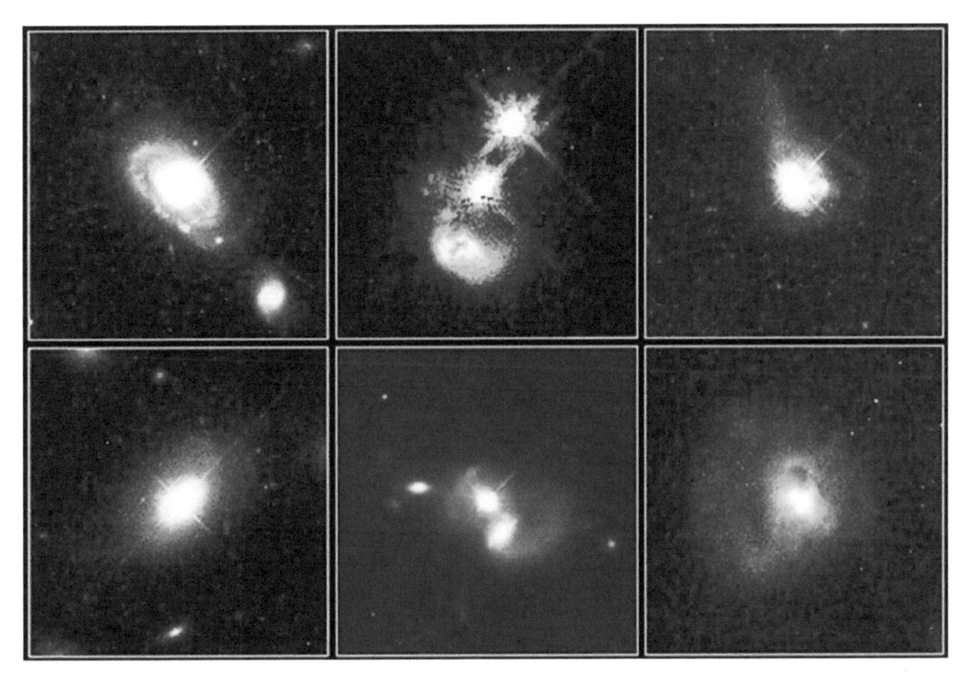

Figure 8.2. Images of six quasars taken with the *Hubble* by John Bahcall and collaborators. The luminous quasars are at the centers of each panel, and are normally so much brighter than their host galaxies that the galaxies are extremely difficult to detect. The brightness of the galaxies has been greatly enhanced here to reveal their detail [credit: STScI/J. Bahcall & M. Disney].

reduction process his team was using to tease out the faint structural details of the host galaxies of quasars. Our conversation continued in my office that afternoon, as John asked for more information about our plans for the HDF. At the end, he became very serious, acknowledging that much thought had clearly gone into our planning—but stressing his view that the HDF was a very risky observation that posed a serious threat to the future of the *HST*. John agreed that various galaxies were sure to be observed in any random field of the sky, but it was quite possible none of them would have high redshift or very large distance. If there were only one or two galaxies at large distances, the huge HDF effort would not be breaking meaningful new ground. John showed me the image of one of the quasar-galaxy objects his team had already observed, which had a known redshift of 0.6, corresponding to a distance and therefore lookback time of six billion (light) years. Given his image of this galaxy, shown in the upper left panel of Figure 8.2, and Mark Dickinson's image of the galaxy 3C 324 at nine billion light years distance, John asked a pointed question: wasn't I concerned that galaxies having distances greater than 6–9 billion light years might not be present in the HDF exposures, in which case the campaign would be seen as a huge waste of precious *HST* time? The *HST*, which had just been repaired at huge expense, would face another enormous backlash.

The HDF was an observational project that had not been subjected to peer review, and it would be using a record number of orbits for the telescope; it could therefore be seen as the whim of a few scientists internal to the Institute. An observation of this type was inevitable at some point in the future with some facility. However, it was the potential timing of an unsuccessful HDF, coming so soon after the name *Hubble* had become a synonym for "trouble," that could end up damaging the *HST* Project and creating a negative aura in Congress and federal funding agencies, not to mention the public, all of which might seriously set back space astronomy. This was John Bahcall's concern.

John's concerns were legitimate, to be sure. It was an important matter of different perspectives. I responded to all of John's points, believing that the *HST* was the best facility to try to get data from the universe at lookback times greater than nine billion years. At some point, the observation had to be tried. Yes, it was risky, but the option of having DD time used for observations of broad community interest was essential to the march of an observational science, and made more feasible when one individual could take the initiative for undertaking a risky project. It must be said that John comported himself thoughtfully and respectfully throughout this conversation. I later came to find out from John's colleagues that he had extremely strong feelings in opposition to the idea of the HDF; given that, John's comportment was admirable and thoroughly professional. By the end of our conversation, he reluctantly accepted my decision that the HDF was going to take place.

In retrospect, my own notes have reminded me that the June Council meeting at which Lyman Spitzer expressed his concerns about the HDF was followed only two days later by the failure of our two-orbit preliminary exposure of the HDF due to the bad guide star—which caused some consternation among our HDF team. That weekend observation was followed just one day later by my lunch and conversation with Bahcall. By chance, these events caused mid-June to be a seminal time for the development of the HDF. Nonetheless, December stood firm as the date for its execution, even though it did still require an additional exposure to confirm the suitability of the new guide star at the slightly revised blank field position of the HDF.

Chapter 9

Acquiring and Releasing the Data

On 1995 December 18, the *Hubble*'s slowly moving Continuous Viewing Zone began its passage over a small spot in the sky, within the constellation of Ursa Major, that we had selected for the Hubble Deep Field. The HDF field would remain within the CVZ for 10 consecutive days before exiting, during which time the telescope would take exposure after exposure of the same field in the four chosen filters. The ultraviolet filter would be used in "bright time" when the *HST* was over the daylit Earth; the other three filters would be used in "dark time" during the passage of the telescope over the night side of the Earth. The goal of the long series of exposures was simple: by electronically combining (or stacking) all of the images, we planned to detect as many galaxies as possible within the small field. How many might we detect? We had no idea, although our rough calculations indicated we should see some dozens of them at a minimum. Our knowledge of galaxy evolution in the early universe was rather uncertain, so we were prepared for most any result.

Preparation for the HDF observing campaign had been going on in earnest for some months, particularly since the late March HDF Advisory Panel had agreed that some effort should be given to undertake a major program to detect distant galaxies. Most of the details on how the Institute carried out the planning and execution of the Deep Field Project is documented in notes and e-mail messages I kept; they record our decisions and actions up to the time of the public release of the fully reduced dataset in January, a mere seventeen days after the last observations.

Most of the initial effort in moving ahead with the Project involved the six Institute scientists who had been the most active in our morning coffee conversations, and who constituted the core group: Mark Dickinson, Harry Ferguson, Mauro Giavalisco, Andy Fruchter, Marc Postman, and myself. We talked constantly among ourselves, tossing ideas back and forth, writing flurries of e-mails to each other and seeking the advice of other scientists who were involved in galaxy research. As various plans were developed for the intricate scheduling of observations, our team recruited other colleagues at the Institute whose expertise was needed

doi:10.1088/978-0-7503-1756-6ch9

to carry out the observations. We did not lose sight of the fact that the HDF was an undertaking done on behalf of the broader astronomical community, and this led us to make two decisions that had a major impact on the complexity of the Project: we would fully reduce the raw data and archive it so an enhanced dataset would be available to all, and we would observe the HDF simultaneously with two *HST* instruments, not just the WFPC2 camera, to produce adjacent deep fields.

The first action would allow non-specialists to analyze the enhanced dataset without having to take on the complex task of calibrating the raw data and removing instrument signatures—a difficult and time-consuming task that Institute scientists could do best. The second decision would allow a separate field to be imaged in another wavelength region. It was possible to use the infrared NICMOS instrument to observe an area of the sky while simultaneous optical imaging was obtained with the WFPC2 camera on an adjacent field. Exposures of two separate fields could therefore be taken at the same time, and implementing both of the above strategies would give added value to the HDF as a research dataset. The infrared deep field could not attain the depth or breadth of the primary visible WFPC2 deep field, but it could—and did—provide significant results for our understanding of the IR universe. The simultaneous operation of two *HST* instruments did entail the need for much more work. It rapidly became clear that we needed to ramp up the HDF team into full campaign mode by augmenting it with Institute staff having the necessary instrument and data expertise to help us put together everything needed to make the observations and handle the complicated telescope scheduling and data acquisition.

Daily planning discussions constantly brought new tasks to light, resulting in our team accreting necessary talent to the group. We ended up with a team of seventeen scientists and technical specialists, fifteen of whom were at the Institute or Johns Hopkins, in addition to the expertise of two European Space Agency scientists. In retrospect, the intense work of the team in such a short period of time was truly remarkable. In nine weeks, they defined the major aspects of the HDF program, which would consist of one primary telescope pointing within the CVZ. Selection of the field required some effort to determine the best location that satisfied the necessary constraints among a number of candidate fields. The team had identified the best WFPC2 filters to be used in isolating four wavelength bands that would provide essential information about the galaxies to be observed in the field. In what turned out to be among the most complex tasks of the HDF campaign, they then developed a detailed procedure, the *data pipeline*, to take the raw data from the WFPC2 exposures and refine them into fully reduced data for the *HST* archive.

By early December, the detailed scheduling of every exposure during the 10 day CVZ passage was well underway. The data reduction pipeline, involving a large amount of software, was being written to correct instrument signatures and apply calibrations for intensity and position so the enhanced dataset could be produced. Computer algorithms were developed that would take the reduced data and apply criteria that defined what objects should be classified as galaxies in order to create a catalog of those galaxies detected by the HDF. The catalog would specify each object observed, its location, and its brightness in the different wavelength intervals—information that would be crucial to follow-up research on the HDF.

The rapid progress on all these fronts was made possible by an energetic team of newly minted young scientists and technical staff, most of whom lacked long-term appointments at STScI. We asked team member Harry Ferguson to coordinate the overall work of the group and organize the completion of important tasks because his experience in scientific and technical issues related to the *HST* had caused him to be a dependable go-to person in most discussions. Fortunately for us, Harry was willing to take on the leadership responsibility for organizing and coordinating most of the work required of the HDF team. Harry had completed his PhD work at Johns Hopkins University three years previously, and following a postdoc research appointment at Cambridge University, had been awarded the prestigious Hubble Fellowship at the Institute. Harry was the right complement to the core HDF group, who were interested in remote galaxies. Focused and knowledgeable about extragalactic research, he was also conversant on most topics and familiar with the characteristics of the *HST* and its instruments. He was always well-organized, and most times when questions arose involving instrument details and data reduction, Harry either had answers available or he knew where to find them. Most importantly, his affable demeanor, coupled with a thorough knowledge of extragalactic research, was such that he easily fit into any group. As the time of the HDF observations approached, the organization and integration of the different team efforts became increasingly critical. Harry took on the role of dealing personally with team members, and in meetings he brought the group's attention to every issue that needed it. His consummate efforts in bringing the HDF to fruition cannot be overstated.

On December 6, less than two weeks before the HDF observations were to begin, the following e-mail message was circulated to the HDF team by another of our scientific and technical leaders, Andy Fruchter, illustrating the fact that important responsibilities were still being defined:

"Dear HDF'ers,

A subgroup met [today] to discuss data reduction schemes and schedules. Below I put forward a suggested list of assignments:

1. *List all exposures, time of exposures, name dataset – Brett Blacker*
2. *Data collection, bookkeeping, file organization – Inge Heyer*
3. *Pipeline calibration – Van Dixon*
4. *Cosmic ray rejection – Mark Dickinson*
5. *Hot pixel identification using darks – Harry Ferguson*
6. *Hot pixel identification using F300W filter – Ron Gilliland*
7. *Registration and co-addition, reduction coord. – Andy Fruchter*

This list of assignments insures that the most important work can be done.

Cheers.

Andy"

All of the above tasks needed to be performed on each of the 343 individual exposures that made up the HDF. What is remarkable is that, in the present era when well-orchestrated research projects are mapped out long in advance in order to comply with the multiple design reviews that are normally required to demonstrate readiness of the project for its continuance, the HDF team was able to organize itself productively in such a short period of time. Only twelve days after the above message was sent out, the team was ready when observations started on December 18. Because all details of the exposures were planned in advance, the ten days of observations went rather smoothly, with the data being downloaded from the telescope via Goddard Space Flight Center to the Institute roughly five times each day. The routine operation of the *Hubble* did not require any of the team to actually be present during a download, but there were frequent occasions when team members would commandeer one of the SUN desktop workstations to take a look at the data as they came in (Figure 9.1).

Much of the work during the 10 day observing period consisted of data organization, e.g., vectoring the data to the right individuals to cache before starting to work on the reduction. During this time, there was a constant flurry of activity, especially because of our decision to finish all data reduction by the time the dataset was to be made public at the annual winter meeting of the American Astronomical Society in January, less than one month away. A number of the more critical aspects of the observations and data handling centered on new techniques that had been developed by team members to enhance the quality of the final HDF image.

During the months preceding the actual open shutter period in late December, the HDF team had undertaken several projects to schedule observations in a way that would produce the best possible dataset. The optimization of the exposures took the form of several procedures, one of which was standard for astronomers using CCDs as detectors, called *dithering*. The other was a new technique developed by team members Andy Fruchter and Richard Hook, which they termed *drizzling,* that

Figure 9.1. HDF team members Ray Lucas, Richard Hook, Harry Ferguson, Marc Postman, and Hans-Martin Adorf examining HDF exposure frames as they were downloaded from the telescope to the Institute [credit: STScI/R. Lucas].

would be important in revealing the structure of faint galaxies by improving image resolution. Both strangely named procedures utilized the characteristics of CCD detectors to extract as much detailed information as possible out of the data. In particular, they were very important in showing structure in many of the small, faint galaxies that were eventually detected in the Deep Field.

Dithering and drizzling make use of the fact that light falling on any part of an individual pixel of a CCD illuminates the entire pixel, which actually diffuses the precise location of the source of the light. For example, consider the image of a star or galaxy that falls near the edge of a pixel. The resulting image indicates only that it was located somewhere in the sky covered by that pixel. However, if another exposure is obtained of the same area of sky by moving the telescope in a slightly different direction by just a fraction of a pixel, that subsequent exposure may find the same star or galaxy falling on the pixel adjacent to the one it occupied in the previous exposure. In this case, a more accurate position of the source can be deduced because it would indicate that the object's position was near the borders of the adjacent pixels. The process of taking successive exposures of a region of sky in which the pointing of the telescope is purposely changed by small amounts corresponding to fractions of a pixel width is called *dithering*. It serves to shift the position of objects to slightly different pixels on the CCD in each exposure. The final image is obtained from these individually dithered exposures by electronically combining them after correcting for the small shifts in telescope pointing. The final combined image can have a spatial resolution that is finer than the size of the detector pixels, revealing smaller detail than would otherwise be possible.

Dithering serves another important function that is valuable in acquiring good images with CCD detectors. Like cameras in common cell phones, every modern astronomical CCD detector contains tens of millions of individual electronic pixels whose sensitivity to light varies from pixel to pixel. Non-uniformities result from the process that produces the silicon layer of the CCD. Certain pixels, referred to as *hot* pixels, are much more sensitive to light and internal electronic noise than neighboring pixels, and each exposure shows these pixels as anomalous bright spots. The opposite holds true for insensitive *cold* pixels that appear as darker spots compared with neighboring pixels. It is essential for astronomical measurements of the intensity of light, which are crucial to understanding celestial sources, to take into account the non-uniform sensitivity of the millions of pixels when interpreting data from the detector. Dithering enables these corrections to be made by means of a simple statistical procedure.

Dithered exposures cause all objects in the frame to be imaged on different pixels in each individual exposure, as shown in the series of three exposures in Figure 9.2. The figure shows a progression of three separate exposures that image the same stars but have their positions shifted slightly. The CCD detector has two hot pixels that appear red. In order to produce the final image from the three exposures electronically stacked together, each exposure is back-shifted by the dithered amount in order to properly align the objects. An individual pixel will therefore occupy different positions with respect to the fixed stars and galaxies in the combined exposure frame. This causes hot and cold pixels to shift around in the succession of dithered images

Figure 9.2. A sequence of three exposures that have been dithered, showing several stars and a faint edge-on galaxy. The different pointing for each exposure is apparent from the shifting positions of the bright stars relative to the CCD's hot pixels, shown as two red spots in the center of the detector [credit: Jerry Lodriguss/AstroPix.com]. Video available at http://iopscience.iop.org/book/978-0-7503-1756-6.

so they can be identified easily and corrected for their anomalous sensitivities by replacing them with the average sensitivity of the other pixels in the dithered exposures that image that same spot of the sky. In this way, pixel variations (or non-uniformities) are averaged out as is shown in the two panels of Figure 9.3. The left panel of the figure shows one of a sequence of dithered exposures that contains numerous cosmetic defects due to cosmic ray hits and abnormal pixels. When the multiple images are combined with the galaxy aligned, the hot pixels are shifted around in the separate dithered images; the final combined, stacked image that results from filtering (or averaging) the dithered exposures is shown in the right panel.

The drizzling process carries the concept of dithering further to improve image resolution, although at the expense of introducing some noise into the image. Drizzling arbitrarily assigns a smaller pixel size to every pixel in each dithered exposure, i.e., it effectively shrinks each pixel to a size that is roughly half that of the actual detector pixel. Then, in aligning the different dithered images by correcting for the dither offsets of each different exposure, the individual dithered exposures are mapped onto arbitrarily smaller pixels that constitute the final combined image, as shown in Figure 9.4. The resulting combined image does have better resolution because of the smaller assigned pixel size that produced the stacked image, so better detail of objects is revealed, as is seen in the example of part of a drizzled image in Figure 9.5. It is important to note, however, that because one never gets something for nothing in the real world, the artifice of confining the signal of an actual pixel to an arbitrary, artificially smaller pixel size can lead to some inaccuracies in the distribution of the light among neighboring pixels in the combined image. This increased noise is the price one pays for achieving the ability to detect smaller detail in the final image of stacked dithered exposures. The dither and drizzle procedures

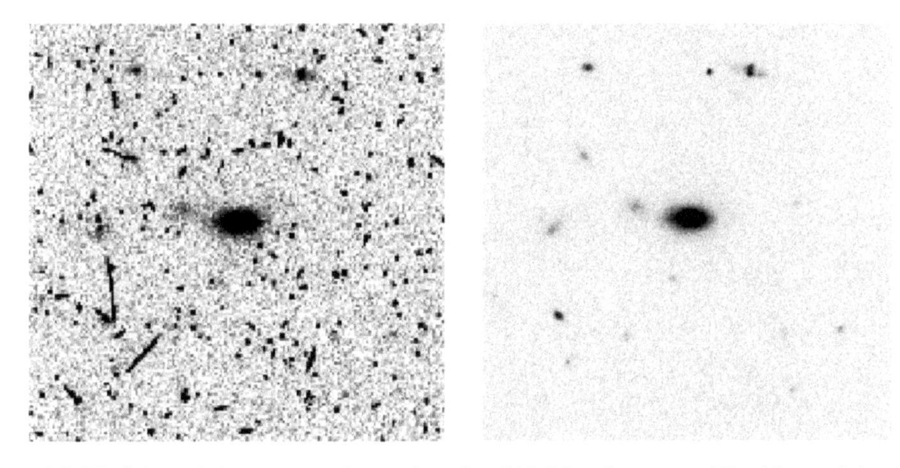

Figure 9.3. The left panel shows one raw frame of a series of 12 dithered exposures. The right panel shows the image that is produced after combining all 12 exposures, discarding any pixel value that is significantly different from the average value of pixels for each location in the sky in the 12 frames. This process filters out bogus signals from hot pixels and cosmic ray hits in individual exposures, producing a clean final image. (Reproduced with permission from Fruchter, A., & Hook, R. N. 1997, Novel image-reconstruction method applied to deep Hubble space telescope images, Proc. SPIE 3164, 120, Applications of Digital Image Processing XX.)

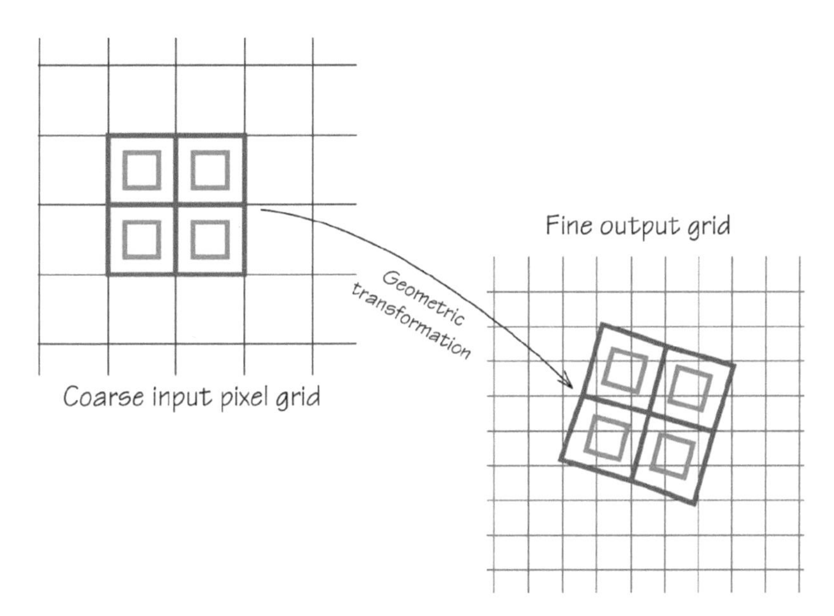

Figure 9.4. Drizzling maps the intensity of each detector pixel onto smaller virtual pixels when stacking dithered images together, thereby creating a higher-resolution image. (Reproduced with permission from Fruchter, A., & Hook, R. N. 1997, Novel image-reconstruction method applied to deep Hubble space telescope images, Proc. SPIE 3164, 120, Applications of Digital Image Processing XX.)

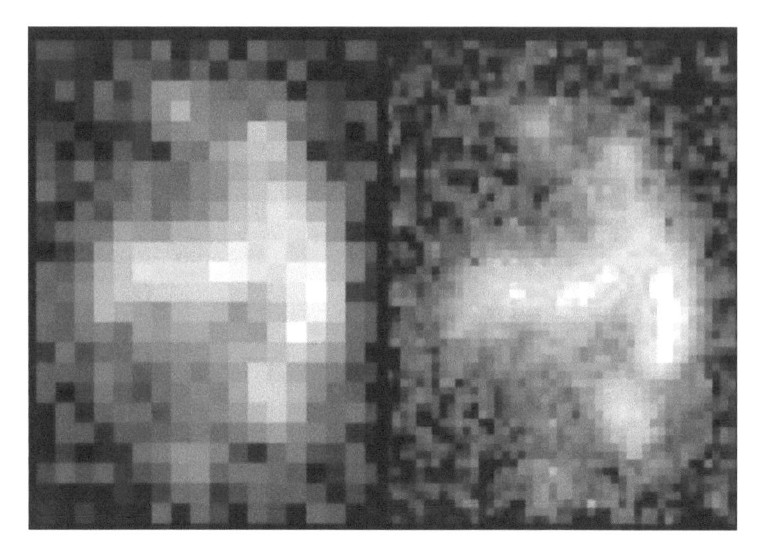

Figure 9.5. An undrizzled image from a series of stacked WFPC2 exposures of a galaxy in the HDF using the original detector pixel grid (left) is compared with the image produced when the exposures are combined using drizzling (right). Finer detail is revealed in the drizzled image [credit: R. Hook & A. Fruchter].

were used extensively in all of the HDF observations and were essential in producing the excellent image quality of the data that was placed in the *HST* archive at STScI and made available to the science community. It has since become a standard procedure for *Hubble* exposures that attempt resolution of fine detail in faint sources.

At no time during the formulation and execution of the HDF campaign did the team disseminate an official announcement of what we were planning, apart from a brief article in the quarterly STScI Newsletter that gave a few specifics. Exacting details of every aspect of the observations and data reductions were eventually described in the final research publication that appeared in a 1996 issue of the *Astronomical Journal*. So many details were being decided upon in the final months before observations were to begin that whatever appeared in writing would have been outdated by the time of publication. Verbal reports on the HDF Project were given to the HST Project at Goddard, to NASA and ESA Headquarters, and to the STScI Council, but the bulk of information about the HDF was passed on via verbal communication between team members and colleagues. Word of mouth information about the Project passed through the professional community and was effective in making astrophysicists aware of the impending observations that were to take place. Based on queries we received from interested colleagues, it was clear there was a keen interest in what an HDF image might reveal. Because the HDF observations were being obtained on behalf of the entire astronomical community, a question was frequently put before us: how would we make the data and the final image available to everyone?

The annual winter meeting of the American Astronomical Society was scheduled to take place in San Antonio in mid-January of 1996, and that seemed to be the perfect occasion to have a media event announcing the completion of the observational campaign and the availability of the HDF dataset. The only problem was that the AAS meeting followed the final HDF observation by a mere seventeen days, requiring a frenzy of data reduction activity if we were going to actually produce the finished product, which included more than 300 individual exposures. We decided to work to the AAS January deadline, and to make a long story short, the team succeeded in doing all the needed work in those seventeen furiously intense days. The team performed heroically, working night and day, including overcoming a 50 year snowstorm that dumped almost three feet of snow on Baltimore, completely shutting down the Institute for several days. It almost, but not quite, came to a "bring in the mattresses" situation so team members could get into and remain in the Institute to do the data handling work. At the same time, we worried about how to keep the computers running when no Institute maintenance staff were available until the snow could be cleared. In a great tribute to a dedicated team, they worked feverishly and produced a superb reduced dataset that has stood the test of more than twenty years' time and rigorous analysis.

The decision was made that we would hold a press conference in San Antonio at the AAS meeting, at which time the Deep Field image would be unveiled and an electronic flag would drop in the *HST* data archive that would permit access to the full HDF dataset by anyone interested in downloading it. Until that time, only team members had access to the data, with the caveat that team members were forbidden to do any science-related analysis of their own with the data. Rather, they were required to confine their work to properly reduce and prepare it for research-level analysis. Team members would have access to do science with the data only at the same time as the rest of the community.

The nuts and bolts work involved in all the data preparation is best suited for more detailed technical reports, so it will not be elaborated upon here. However, we ought to devote some attention to the preparation of the primary visual legacy of the observations—the full color image of the Deep Field. One of the advantages of having obtained exposures of the HDF in four separate wavelength bands, which was done for research purposes to aid in the interpretation of the stellar content of the galaxies detected, was the fact that three or more monochromatic images can be used to construct a realistic color image in the same way that color images on a television screen consist of a combination of closely packed red, blue, and green pixels. A color print is made in virtually the same way by combining ink of different colors, often cyan, magenta, yellow, and black, to produce a faithful color image. There was no doubt that the final HDF product that was likely to produce the most visceral impact was going to be a high-resolution color image of the field.

Over the lifetime of the *HST*, the Institute has devoted a great deal of effort to creating color images of *HST* targets for both scientific and public relations purposes. What better way to begin understanding the universe than to see it as our eyes visualize it? If ever a picture was worth a thousand words, the HDF was certain to be a prime candidate. Color images from the *HST* are the stock and trade of the telescope's connection with the outside world. Few telescopes match the ability of the *Hubble* to show features that haven't been seen before, although there are recent telescopes, both on the ground and in space, which are sensitive to radio and X-ray wavelength regions and give the *HST* good competition in some circumstances. A thorough archive of hundreds of the *Hubble's* most memorable images exists online on the *hubblesite.org* website maintained by the Institute. The large majority of those images have been created from data obtained by astronomers using the *Hubble*, and the task of creating the finished color product resides in a division of the Institute devoted to education and outreach where they have for years been generated by one individual—Zolt Levay.

Arguably the most irreplaceable person at STScI, Zolt has developed his skills in image creation and refinement at the Institute for 30 years. His task has been to take data from the telescope and produce a faithful visual representation of what the instrument has detected. Much of Zolt's work involves some standard procedures, e.g., cleaning of the images from cosmic ray hits, and corrections for hot and cold pixels. Problems with dithered exposures that are not registered correctly, thus causing degradation of the resolution, occasionally need attention. However, one of the most important aspects of creating those *Hubble* images that are admired as examples of the beautiful and strange in the cosmos is achieving the proper color balance for each image.

To a large extent, color balance is subjective—as much art as science. The ability of the eye to discriminate color depends sensitively upon the brightness of the object being observed. Even with the largest telescopes, virtually every astronomical object observed by eye, except the few brightest planets and stars, appears very pale. For most observational astronomers, the true rendition of an astronomical image would appear as if on your television screen with the "color" dialed down near its minimum setting and "contrast" also muted. Such an image, while faithful to astronomers, fails to receive "oohs" and "ahhs" from the public when viewed. Thus, there is a tendency for the media to crank up the color and contrast on most *HST* images that are shown to the public. It is ultimately a judgment call as to what truly represents the appearance of an object that has been imaged by a telescope. In his work with *HST* images, Zolt must pay close attention to getting color balance correct, which is one of the scientific facets of creating a color image. The hues, i.e., red versus blue versus yellow versus green, of his color images are as correct as the data allow from the relative intensities through the different filters, and he resists as much as possible the tendency to "spark" the images by oversaturing color, as many television commercials do.

The color image of the HDF that was created by Zolt from the data obtained with the WFPC2 camera imaging in the different filters is shown in Figure 9.6. It is in many ways the real face of the HDF, although it must be admitted that most of us in the scientific community find it as intriguing an image as we've ever encountered. Most of the information in the image exists in the myriad of small fuzzy objects, the vast majority of which are galaxies. Only nine stars from our own Milky Way galaxy appear in the image as relatively bright foreground objects. The remainder of the objects are galaxies, approximately 2700 in total, and they cover a large range in distance from us. Generally, the larger and brighter galaxies are among the closer objects, while the faint, barely resolved small objects are more likely to be more

Figure 9.6. The full high-resolution color image of the Hubble Deep Field, created from the hundreds of exposures obtained by the *HST* over the 10 day observing period [credit: STScI/Z. Levay]. A larger version of this figure is available at http://iopscience.iop.org/book/978-0-7503-1756-6.

distant galaxies. However, this is not always the case, because many galaxies that appear small and faint may be intrinsically small and faint, and therefore nearby. Therefore, a knowledge of the distance to each galaxy, which also determines how far back in time we are observing it, is essential to gleaning scientific information from the image. Without a knowledge of distances, the Deep Field gives us a wonderful depiction of what space contains but it is barren in terms of what it could reveal about galaxy evolution. Fortunately, we had anticipated this need and were already working to obtain galaxy distances months before the HDF observations were made.

Chapter 10

Galaxy Evolution Revealed

One of the defining discoveries of the 20th century was the demonstration by Georges Lemaître and Edwin Hubble, using spectra obtained by Vesto Slipher, that the universe is in a state of uniform expansion. Once the expansion rate—normally specified in terms of a parameter called the Hubble constant—is measured, it provides a straightforward way of determining the distance between any two objects. One need only determine the speed with which a galaxy is observed to be moving away from our Milky Way galaxy in order to know its distance. The velocity of an object radiating light can be measured straightforwardly by means of the Doppler effect, simply by obtaining its spectrum.

Early in the process of defining the Deep Field, the team understood that, if any interesting astrophysics were to come out of the observations, it was essential that the distances to galaxies detected in the image be known. Spectra needed to be obtained for as many galaxies in the image as possible, otherwise the HDF would end up being little more than a nice picture. Because the spectrum of an object disperses the light across wavelengths, it distributes the light over many pixels on the spectrograph detector. The net effect of smearing the light of a star or galaxy over many pixels is that the spectrum appears as a much weaker signal on each pixel, compared to the brightness of the direct image of the galaxy. This causes the brightness level of the spectrum of an object to be so much fainter than the direct image of the object that its registration (or imprint) on the detector pixels can easily be dominated by sky background light and the intrinsic electronic noise of the detector.

Distant galaxies are genuinely faint objects in any telescopic exposure, even with the largest telescopes, as the HDF image reveals. Only the brightest galaxies in any image can possibly have spectra taken such that the spectrum is detectable in a reasonable exposure time. The fainter galaxies would require such impractically long exposures and Herculean efforts to produce a detectable signal that astronomers must accept the fact that compromises must be made if spectral information is to be

doi:10.1088/978-0-7503-1756-6ch10

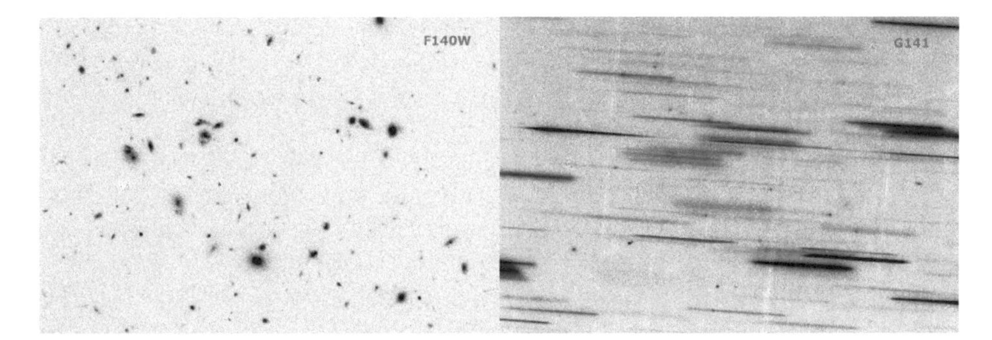

Figure 10.1. The left panel shows a direct *HST* image of a distant cluster of galaxies. In the right panel, the spectra of the cluster galaxies are imaged with a prism placed in the telescope beam, providing the redshift and other information about the galaxies. Because the spectra are disbursed over many more pixels than the direct images, the image of the spectra requires 20 times longer exposure time than the direct image on the left in order to achieve the same acceptable level of brightness for analysis [credit: Ben Weiner/Steward Observatory].

obtained. A good illustration of this effect is shown in Figure 10.1, where the two panels show the direct image of a small cluster of galaxies compared with an image of the spectra of those galaxies. The exposure time for the image of the spectra displayed in the right panel of the figure, which are visible for only the brighter objects in the left-hand panel, was almost 20 times longer than the exposure time for the direct image. The spectra of the fainter galaxies in the left-hand image do not even register in the right-hand frame.

As soon as the position of the HDF had been selected, we contacted astronomers who had access to what was at that time the largest ground-based optical telescope in the world, the Keck Observatory 10 m telescope (see Figure 1.4) on the island of Hawaii. We advised them of the upcoming HDF campaign and the need to obtain spectra of the brighter galaxies in the image. The Keck telescope is operated by the consortium of Caltech and the Universities of California and Hawaii. Fortunately, a group of astronomers at those universities were keenly interested in the HDF and were willing to devote a significant amount of their personal observing time on the Keck telescope to obtain spectra of galaxies in the HDF as a means of determining their distances. Due to the time that would be needed to get the spectra of HDF galaxies—which were guaranteed to be quite faint, given the absence of any bright objects in the Palomar Sky Survey photo of the Deep Field region—the Keck astronomers were anxious to get started in taking spectra as soon as possible.

The HDF team was equally eager to jump-start the acquisition of spectral data, because the entire HDF Project would have little significance without that information. Therefore, months before the December HDF observations were to take place, the team had made the decision to take the preliminary two-orbit exposure of the Deep Field with the *HST*, mentioned previously, in order to detect the brightest galaxies in the field. This image would be provided to the Keck observers and they could use it as a finding chart to get a head start on obtaining spectra of those galaxies. The image did show roughly 70 galaxies that appeared bright enough for the Keck 10 m to attempt to get their spectra, from which the galaxy velocities could be measured. The Keck observers, including Len Cowie, Judy Cohen, David Koo, and colleagues, immediately began acquiring spectra and determining distances to the galaxies, which they made non-proprietary by posting them on a website with public access.

Before the HDF observations began in December, the Keck observers had already obtained around 40 spectra of galaxies that appeared in the initial preliminary two-orbit image. The full 150 orbit HDF image eventually revealed more galaxies that were sufficiently bright that the acquisition of their spectra could be attempted. Within 18 months of the HDF observations, the Keck astronomers had obtained spectra of 125 of the brighter galaxies in the HDF, from which their velocities and distances were determined. These spectra transformed the HDF image into an important tool for understanding the evolution of galaxies and the nature of the early universe.

The expansion of the universe causes features in the spectra of galaxies to be shifted to longer, i.e., redder, wavelengths. The amount of the redshift, as it is called, is different for every galaxy and is directly related to its recession velocity, which in turn is determined by the galaxy's distance from us. The fact that light has a finite velocity means that we observe all objects in the universe as they were when the light left that object. The time required for the light from each galaxy to reach our telescopes is called the lookback time for the galaxy. Thus, a galaxy's distance, its recession speed, and its lookback time are all related to each other. Astronomy is unique in that it permits us to see directly into the past over significant timescales.

The redshifts of galaxies have historically been used by astronomers as a convenient expression for distance and lookback time, when applied to distant objects such as galaxies and quasars. It is the one attribute that describes the place of a galaxy in constructing the evolution of the large-scale structure of the universe. In any discussion of galaxies other than our own Milky Way, one of the first questions any presenter of galaxy data will be asked is what the galaxy's redshift is. No other attribute is as important to determining how far away the object is and at what epoch we are seeing it in the timeline of the history of the universe.

The concept of redshift is easy to understand if one grasps the Doppler principle: the observed wavelength of a propagating wave is related to the velocity of the object with respect to the observer. The radiation from receding galaxies in the expanding universe clearly experiences a shift in its observed wavelength, as depicted in Figure 10.2. The redshift is defined as the amount of the wavelength shift from its initial value at rest to the value we observe with our instruments, divided by the rest wavelength. If the shift of some spectral feature with a rest (laboratory) wavelength of 5000 Å is observed to be 50 Å to the red because of the outward velocity of the galaxy, its redshift is 50/5000=0.01.

Galaxies that have greater distances from us—and are therefore seen at earlier epochs in the universe—have higher redshifts. The redshift of a galaxy is easy to measure as long as it is bright enough to have its spectrum taken. There is a straightforward relationship, derived from observational and theoretical studies, between a galaxy's redshift and the lookback time at which we are observing it. The relation is fundamental to extragalactic astronomy, having been determined from

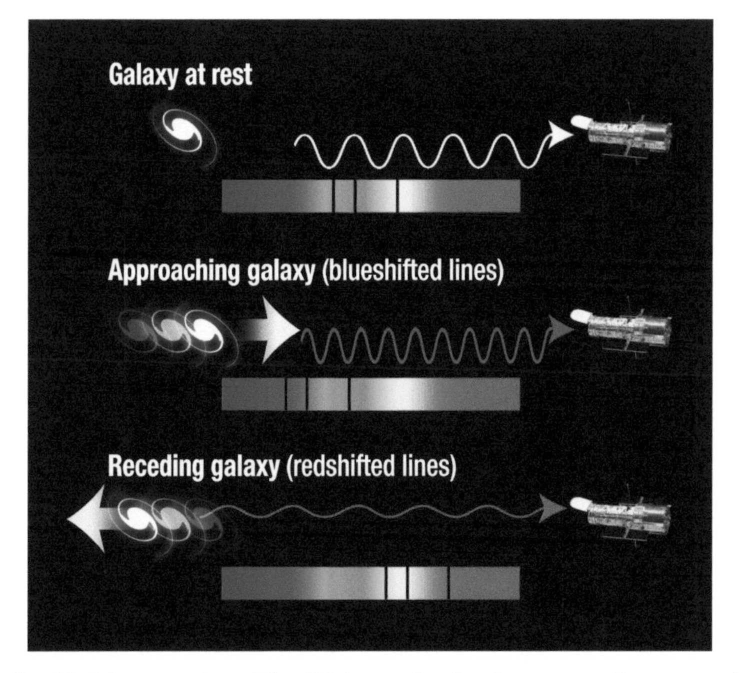

Figure 10.2. Graphical demonstration of the shift in wavelengths of a spectrum that occurs when a galaxy moves toward or away from the observer. The shift in the spectra of more distant galaxies, which have larger recession velocities, steadily increases such that, for redshifts of $z > 10$, most of the radiation is shifted into the infra-red region [credit: STScI/A. Feild].

the known expansion rate of the universe and current cosmological models, and is shown as a plot in Figure 10.3. Higher redshifts represent increasingly longer lookback times as one advances back toward the epoch of the creation event of the present universe, 13.7 billion years ago. However, the relationship is not linear. A galaxy of redshift $z = 1$ (redshift is traditionally represented mathematically by a lowercase z) represents a lookback time of almost eight billion years, more than half the age of the universe since the Big Bang. A galaxy observed with redshift $z = 10$, near the limit of the most distant galaxies yet observed with the *Hubble*, has a lookback time of 13 billion years.

Shortly after the HDF observations had been completed and the dataset made available to the community, HDF team members assembled all the redshifts of the HDF galaxies that had been measured and posted by the Keck observers at that time, amounting to about 90 of the 125 total that would eventually be obtained. In order to indicate the appearance of galaxies as a function of redshift, and therefore lookback time, the redshifts were displayed in the HDF image next to the corresponding galaxies. The annotated image is shown in Figure 10.4. It presents a clear representation of what galaxies looked like at different epochs as the universe evolved in its expansion from the Big Bang. The nearest galaxies that were imaged in the HDF have redshift $z = 0.09$, with a lookback time of one billion years, and the most distant galaxy with a measured redshift has $z = 4.02$, observed at a lookback time of 12 billion years ago, i.e., 1.7 billion years after the Big Bang. The HDF is

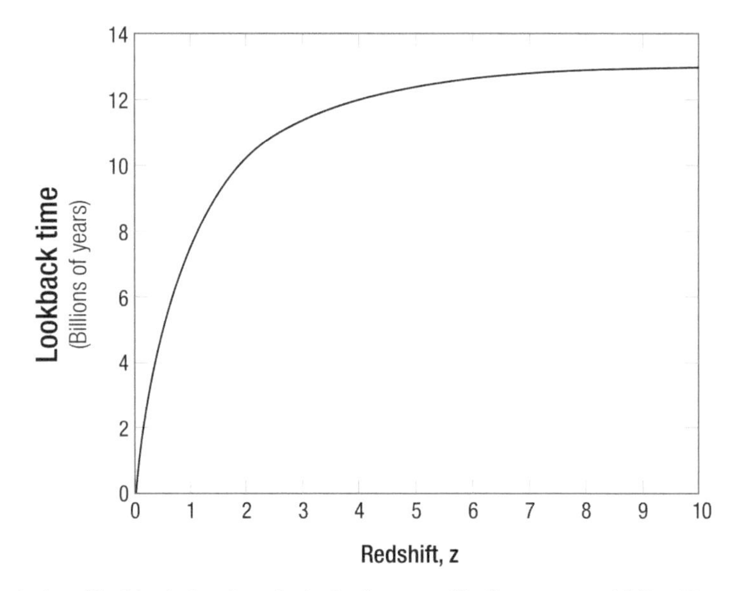

Figure 10.3. A plot of lookback time for galaxies having a specific distance, or redshift z. For $z = 1$, we see a galaxy as it was roughly 7–8 billion years ago, more than half the age of the universe. For a redshift of 10, we see a galaxy as it was 13 billion years ago, only 700 million years after the Big Bang [credit: STScI/A. Feild].

therefore sampling galaxies over a time interval that represents 90% of the history of the cosmos since the Big Bang creation event.

Studying the HDF image with redshifts annotated next to the galaxies in Figure 10.4 is compelling for any astronomer. It is the astronomical counterpart to an archeologist digging into layers of the Earth to reveal artifacts of the past, except in our case we are actually observing these galaxies as they existed long ago, not as artifacts. Not surprisingly, the lower-redshift, nearby galaxies appear larger and brighter than the very small, quite faint, extremely distant high-redshift objects. There is a clear difference in the morphology of the galaxies over the range of redshifts. This evolution of galaxies over time is more evident in the different format utilized in Figure 10.5, which shows, in a more systematic manner, the changing morphology for different lookback times of HDF galaxies whose spectra were obtained with the Keck telescope.

Figure 10.5 was assembled after the majority of the galaxy spectra accessible by the Keck telescope had been acquired. It shows 110 galaxies from the HDF whose redshifts were measured, with the galaxies all displayed to the same relative scale. The galaxies are positioned in the diagram according to two factors: the first is their redshift, i.e., their distance and therefore lookback time, which increases toward the bottom; the second factor is their intrinsic brightness, or luminosity, which increases to the right. Starting at the lower right of the figure and moving upward to the left, the galaxies form a sequence that represents a flow in cosmic time and intrinsic brightness.

The fact that galaxies occur in an apparent diagonal strip in Figure 10.5 is an observational selection effect. The absence of galaxies in the lower left part of the

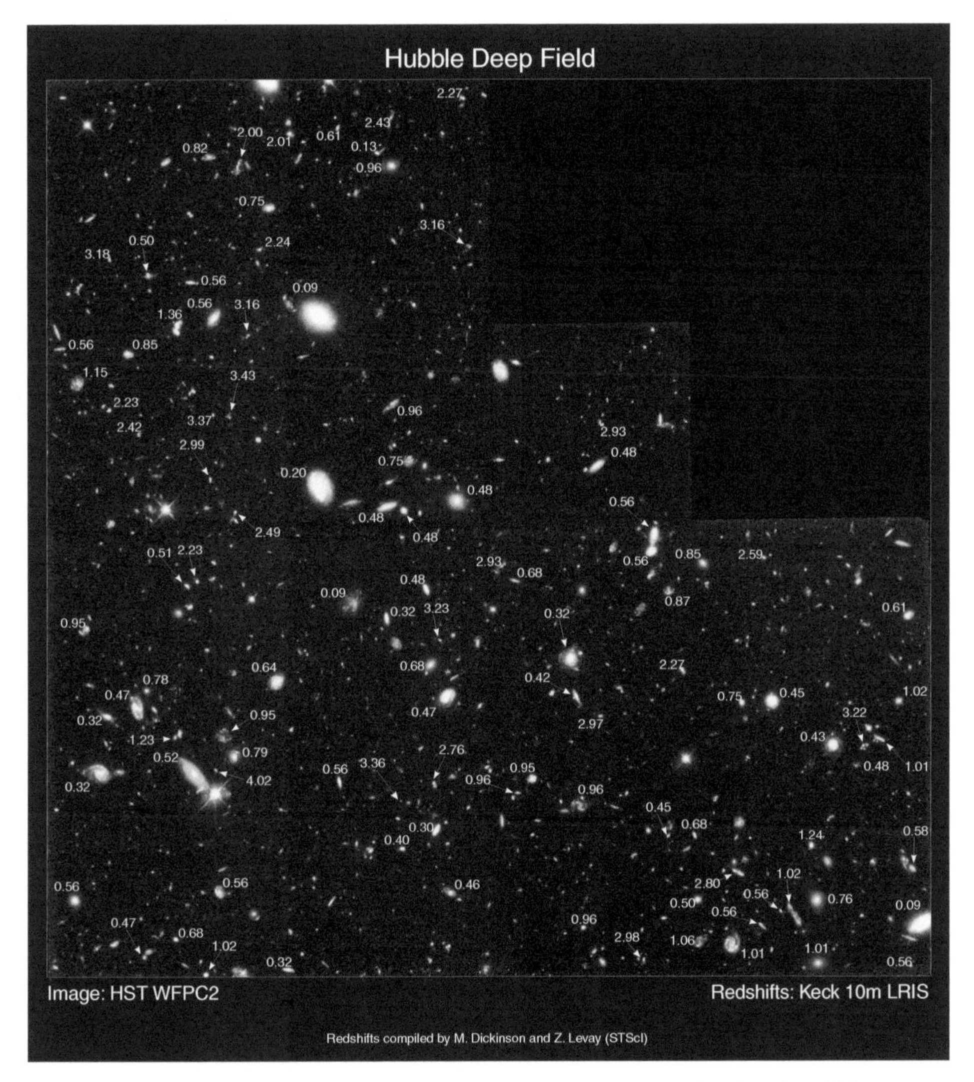

Figure 10.4. The Deep Field image with the Keck redshift placed next to each galaxy for which a spectrum could be obtained. The redshift provides the distance and lookback time to the galaxy [credit: M. Dickinson & Z. Levay]. A larger version of this figure is available at http://iopscience.iop.org/book/978-0-7503-1756-6.

diagram, which represents very distant galaxies of low luminosity, is due entirely to the fact that such galaxies are too faint to be detected, even with the *Hubble*. Those galaxies may exist in abundance, but they would all be below our current detection limits. The paucity of galaxies in the upper right portion of the diagram is due to the fact that very luminous galaxies are probably so rare that the HDF did not sample a sufficiently large area of the sky to detect any.

By following the mosaic of galaxies in Figure 10.5 from the bottom to the top, one can see how galaxies have evolved over time, and such evolution is revealed to be quite clear. The high-redshift galaxies all have a chaotic appearance; only at lower

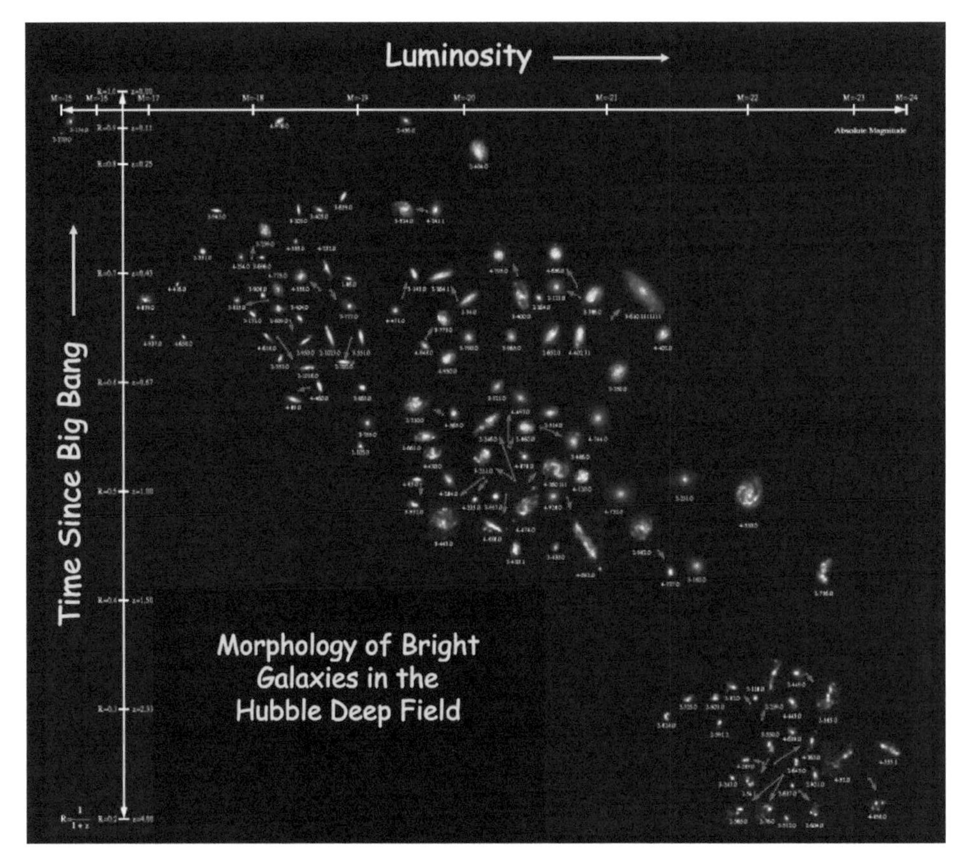

Figure 10.5. HDF galaxies with measured redshifts are placed according to their distance (lookback time) and intrinsic brightness (luminosity). The image of each galaxy is displayed on a scale that shows the relative sizes and morphologies of galaxies at different lookback times [credit: Y. Park/R. Allen/H. Ferguson/N. Panagia]. A larger version of this figure is available at http://iopscience.iop.org/book/978-0-7503-1756-6.

redshifts do galaxies appear more symmetrical, with an ordered structure. This difference is made evident from the enlarged images of a number of the distant, high-redshift HDF galaxies that occupy the lower right portion of Figure 10.5, which are presented in a mosaic in Figure 10.6. In the early universe, none of these galaxies appear anything like the Milky Way or local neighborhood galaxies such as those shown in Figure 6.3.

Also notable is the fact that the galaxies in the lower right part of the diagram are small and blueish in color. The blue hue comes from the dominance of massive, hot young stars that are apparently forming at prodigious rates in galaxies in the early universe. As stars become older, slowly but steadily depleting the hydrogen fuel that energizes them via nuclear fusion reactions, the stars react in different ways depending upon their mass. Many expand and develop cooler outer layers, thus taking on a reddish appearance. Others collapse, explode and die out. In time much of the gas out of which new stars form is steadily used up, so the star formation process ramps down in time. Cooler stars radiate more effectively in redder portions

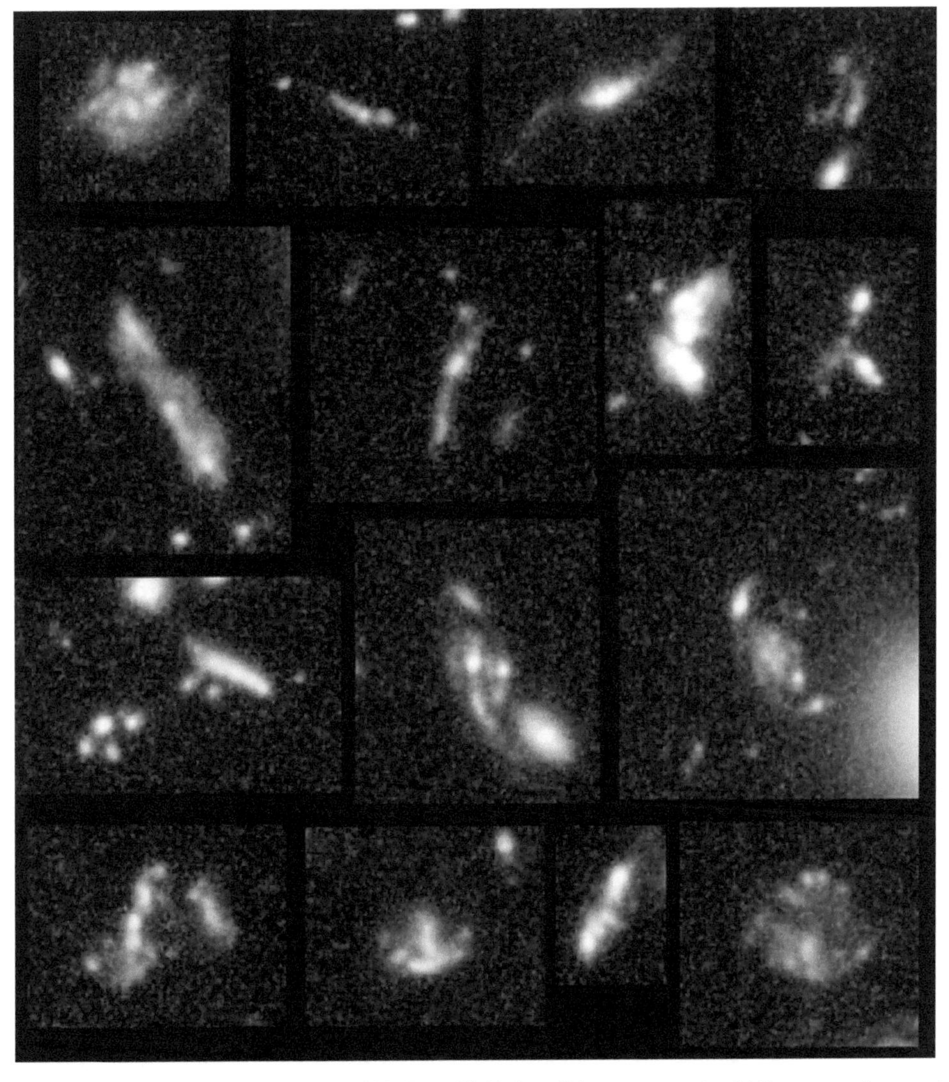

Figure 10.6. Details of galaxies in the Hubble Deep Field that all have measured redshifts greater than $z = 1$, corresponding to distances exceeding seven billion light years [credit: M. Dickinson].

of the spectrum. Hence, there is a tendency for groups of stars in galaxies to take on a blueish color if they have significant number of young stars that have formed recently. The change in galaxy color from blueish to reddish as one goes from large to small lookback times indicates how galaxies age and evolve. It shows clearly that the formation of new stars occurred in galaxies much more actively in past epochs than at the present time, 5–10 billion years later.

The Figure 10.5 mosaic represents one of the more important results of the Hubble Deep Field. In the lower part of the mosaic, high-redshift galaxies are clearly much smaller and more irregularly shaped, i.e., less symmetrical, than present-epoch, low-redshift galaxies in the upper part of the mosaic. Following the flow of galaxies in time from lower right to upper left in the figure, a steady evolution is clear in which galaxies increase in size and symmetry with time. Detailed computer simulations that model the gravitational interaction of galaxies with their surrounding environment over time confirm that accretion of gas onto galaxies and the mergers of galaxies are common processes that occur frequently and can explain the evolution seen in the mosaic of galaxies in the figure. Both the HDF image and theoretical model calculations support the idea that galaxy formation and evolution is largely a "bottom-up" process that starts with the collapse and fragmentation of pockets of gas from self-gravity in the initially expanding primordial gas created at the time of the Big Bang, resulting in the formation of groups of "seed" galaxies that are initially small in size, relative to current-epoch galaxies.

Many details of our current understanding of galaxy evolution have come from complex hydrodynamic calculations performed on supercomputers that apply known physics to the environment that we observe in the early universe from observations of galaxies over a wide range of conditions and cosmic epochs. A number of these studies present their results in video form, showing in striking visual detail how galaxies evolve over time in their size, structure, and stellar composition as a result of ongoing interactions. The external interactions are primarily collisions and mergers with other galaxies and large gas clouds, brought about by their strong mutual gravitational attraction. Internal processes within the galaxies, related to the evolution of each galaxy's own stars and interstellar gas, also play a role.

Because galaxies consist of many millions of stars, the correct physical treatment of the processes associated with a galaxy requires the myriad of stars to be represented in a realistic way. This is done through *many body* or *N-body* simulations whereby the numerous stars in the galaxy are represented by points that have mass and motion in order to treat their gravitational energy and velocities, as well as their winds and ejecta if they go into outburst. Model simulations can require months of supercomputer time, but the ability of such calculations to follow the history of galaxies over time has transformed our understanding of how the largest objects in the cosmos evolve.

The results of several well-known studies of galaxy evolution are presented here in two videos that are instructive in showing how early galaxies are believed to form and evolve. The first video, shown in Figure 10.7 (readers of the printed version of this book can also access it online) was produced by researchers at the Universities of Zurich and California; it shows a representative region of the universe as it evolves shortly after the Big Bang. The upper right corner of the video displays the time elapsed following the Big Bang in terms of the redshift, z. The video begins at a time corresponding to redshift $z = 90$, roughly one-half million years after the Big Bang, when the expanding universe still consists of very uniform, hot glowing gas. Shortly thereafter, expansion cools the gas to a temperature where it stops radiating, at $z \approx 15$, when the video background becomes dark and significant gravitational condensations begin to form. Substantial filaments of gas having masses corresponding to those of galaxies occur at redshifts around $z = 10$, which corresponds on the Figure 10.3 graph to a time 13 billion years ago—roughly 700 million years after the Big Bang. Like bees in a swarm, these initially formless clumps and filaments possess

Figure 10.7. A supercomputer simulation of the condensation of gaseous filaments in the expanding universe after the Big Bang that depicts how random clumps of gas accrete together to form a large swirling mass that eventually becomes a spiral galaxy [credit: P. Madau/U. Zurich]. Video available at http://iopscience.iop.org/book/978-0-7503-1756-6.

their own velocities from the universal expansion, and their mutual gravitational attraction causes them to interact. The transfer of their momenta as separate entities into one combined object results in a rotational motion when clumps merge together. This rotational motion causes the accreting galaxy gas to flatten into a pinwheel spiral shape.

The Figure 10.7 video realistically depicts how a formless protogalaxy filament of gas in the very early universe interacts with other nearby, passing gas clouds. It represents how the first galaxies in the universe can form out of the expanding formless gas via the coalescence of initial gas clouds. The simulation follows their gravitational and gaseous fluid interactions as diffuse clouds progressively collide and accrete. An initially relatively small, formless, gaseous filament slowly grows in size and develops a rotational velocity that it acquires from the momentum transferred by the accreted passing clouds. Over a period of some hundreds of millions of years, the protogalaxy ends up larger and takes on a symmetrical spiral structure similar to that of the Milky Way galaxy.

As is true of most complex numerical calculations, the simulation depends on some assumptions about the conditions surrounding the galaxy during its evolution. In this video, the consequence of significant mergers with other galaxies of comparable size and mass would influence the spiral nature of the galaxy. Such mergers are expected during the life of a galaxy and may be the most important

Figure 10.8. A simulation of two well-developed spiral galaxies, gravitationally bound and interacting in a merger that fundamentally changes their structure from their initial spiral shapes to a larger, much more diffuse structure that has similarities to an elliptical galaxy. (Reproduced with permission from J. Dubinski/CITA/U. Toronto) Video available at http://iopscience.iop.org/book/978-0-7503-1756-6.

aspect of determining the morphology of galaxies, even to the extent of removing the spiral nature of the galaxy, as some of the Hubble deep field surveys following the original HDF have observed. That said, the early spiral shape and rotation of the galaxy are believed to be realistic in showing how many galaxies begin their existence.

The second video is a many-body calculation representing two interacting spiral galaxies at evolutionary phases later, i.e., at lower redshift and a more recent cosmic epoch, than the early forming galaxy of the Figure 10.7 video. In this sense, the second video illustrates just how mergers serve to strongly shape galaxies. Two well-formed spiral galaxies approach each other and become gravitationally bound together, so they begin merging, as shown in Figure 10.8. They slowly merge via a series of passages around and through each other, to eventually form a single, larger galaxy that no longer retains its previous shape as a spiral galaxy. Rather, the merger produces a spheroidal (or ellipsoidal) galaxy lacking the more ordered structure of its two predecessor spiral galaxies. The merger also results in a dense concentration of mass in the center of the galaxy; that mass, in all likelihood, evolves into a black hole. This video demonstrates how the gravitational merger of two spiral galaxies can result in the formation of a larger, more elliptical-looking galaxy bearing little resemblance to the well-defined symmetrical structure of the two predecessor galaxies. From calculations such as those of this video, as well as direct observations with the *HST*, astronomers are confident that galaxies do change their shapes and sizes in this way over cosmic time. The two videos represent very realistic, likely scenarios that explain how most galaxies in the local universe, as shown in Figure 6.3, evolve to their observed present structure.

These videos are fine examples of a number of forefront research efforts demonstrating that, when known laws of physics are applied to realistic conditions in the early universe, the evolution of galaxies can be shown to proceed in a way that conforms to what we observe for the galaxies in the Hubble Deep Field.

Robert Williams

Chapter 11

Star Formation History and Photometric Redshifts

In early team discussions about which filters to use in acquiring data, we made the decision to take exposures through an ultraviolet (UV) filter even though the sensitivity of the WFPC2 camera was low in this wavelength region. Although the number of galaxies detected in the UV exposures would be far less than those detected in the more sensitive blue, green, and red longer-wavelength filters, several reasons dictated the choice of the UV filter. First, the *HST* spends half of every orbit passing above the daylit side of the Earth, which causes increased scattered sunlight from the Earth's atmosphere to significantly limit the depth of any exposures taken in the longer-wavelength filters during this time. Exposures using the less sensitive UV filter are much less affected by this problem. Second, because UV wavelengths are blocked by the atmosphere and can only be accessed by telescopes in space, the UV exposures could bring to light previously unknown information about galaxies. In particular, hotter stars tend to radiate much of their light at shorter wavelengths, so the UV image might well reveal galaxies that contain numerous hot stars that are otherwise not identified by ground-based telescopes. Inclusion of the UV was, for us, a well-worthwhile strategy.

The hottest stars are generally also the most massive, and through nuclear fusion reactions, they expend their reservoir of hydrogen fuel much more quickly than the far more populous lower-mass, cooler stars that predominate in galaxies. When astronomers see concentrations of blueish stars, they are seeing young stars that have formed recently, in astronomical terms. Some fundamental cosmological questions have intrigued astronomers for decades: How does the original hot, dense gas that emerged from the explosive expansion of the Big Bang cool and gravitationally collapse to form stars and galaxies? Did this process first occur on grand scales or was it triggered on small scales? Was the cooling condensation process efficient in binding up most of the gas, such that the great majority of stars and galaxies formed shortly after the Big Bang? Or was the process less efficient, leaving

doi:10.1088/978-0-7503-1756-6ch11

tenuous clouds of gas around so star formation would continue occurring over billions of years up to the present time? The answers to these questions would tell us, in many ways, the basic history of the universe.

During the planning of the HDF observations, members of the HDF team, along with their Institute colleague Piero Madau, came to realize the significance of the UV image of the HDF in addressing the question of star formation history. If they could determine rough distances to the galaxies in the HDF, and therefore the lookback time to each, they could compare the brightness of each galaxy in the UV versus the red filters. The relative intensities in the two wavelength intervals would indicate how many stars in that galaxy are young, i.e., blue and recently formed, compared to those that are older, i.e., red. The relative rate of current, ongoing star formation in that galaxy would then be determined. That information could be compared with the rates for galaxies having different lookback times, and the universal history of the rate of star formation could be determined straightforwardly by this process.

The difficulty with the above procedure is that the HDF image shows more than 2500 galaxies, whereas the Keck spectra, for which distances and lookback times were therefore known, existed for only roughly 125 of the galaxies. The fundamental question arose: how might one use the HDF images in the different filters to get a rough idea of the lookback times for all of the large number of galaxies that are detected in the UV filter? Lacking spectra, the Madau group came up with an ingenious way to get approximate distances from the HDF filter images. Hydrogen is by far the most abundant element in the universe and the vast majority of stars, and its atomic structure causes it to absorb strongly at UV wavelengths. Hydrogen produces a sudden drop (or break) in the radiation that stars emit in the ultraviolet. This drop in brightness causes stars and galaxies to appear brighter in exposures taken with the longer-wavelength filters, i.e., visible and red regions, than they appear at shorter wavelengths. This drop in radiated light is shown in Figure 11.1, and is very apparent when one compares the brightness of galaxies in images taken with the different filters.

For distant galaxies that are expanding away from ours at increasingly higher speeds, the Doppler shift causes the hydrogen break in brightness to be shifted to longer wavelengths as the spectrum is redshifted. This is illustrated in Figure 11.1 by the solid black lines that show how the light from a galaxy appears as it is shifted to longer wavelengths for redshifts of $z = 3$, 4, and 5, corresponding to lookback times of 11, 11.5, and 12 billion years. The dotted and dashed lines show the transmission of the four *HST* WFPC2 filters, labelled U, B, V, and I, that were used for the HDF. We see that the observed drop in brightness of a galaxy occurs in different filters depending upon its redshift. By comparing the relative brightness of every HDF galaxy through the different filters, Madau and his collaborators were able to determine approximately where in wavelength the break occurred, so the approximate galaxy redshift could be determined even though its spectrum had not been obtained.

Figure 11.2 shows an example of a sequence of images of five distant galaxies taken in different wavelength filters, labelled V, Y, J, and H, for which crude redshifts of the galaxies can be determined by the technique of the spectrum break.

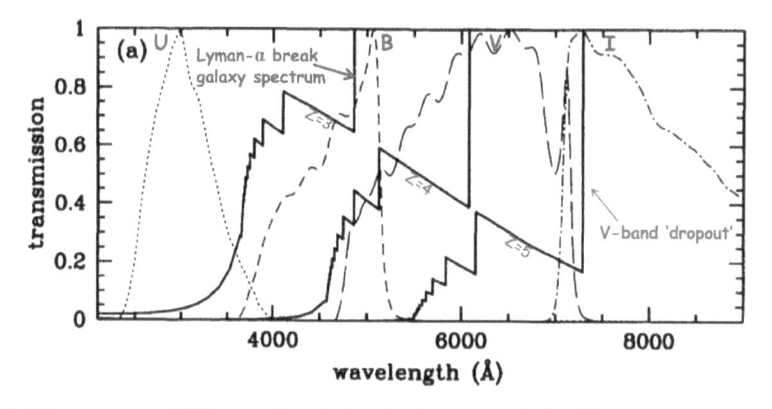

Figure 11.1. The concept of "drop out" galaxies is demonstrated in this diagram, which shows the transmission over wavelength for the four filters U, B, V, and I (dotted/dashed lines) used for the Hubble Deep Field. The solid lines represent the intensities of the spectra that would be observed for galaxies having redshifts of $z = 3$, 4, and 5. The sharp blueward decrease in intensity caused by hydrogen absorption results in such galaxies not being detectable in bluer filters. The redder wavelengths where the galaxies could be detected enable their approximate redshifts to be determined. (Reproduced from Madau, P., et al. 1996, High-redshift galaxies in the Hubble Deep Field: colour selection and star formation history to $z \sim 4$, MNRAS, 283, 1388, by permission of Oxford University Press.)

The panel of galaxies images shows them as small, indistinct blobs of light in exposures taken with the *Hubble* through the four different wavelength filters. The filters transmit light of successively longer wavelengths from the visible into the infrared region of the spectrum. In each case, the galaxies are too faint to be detected in the visible V and Y filters, but they suddenly appear in the longer-wavelength J and H filters. The wavelength of the break in their spectra between the Y and J filters indicates that all of the galaxies must have the hydrogen break at a wavelength between those of the Y and J filters, that is, about 1.1 micron, or 11,000 Å. Given that the hydrogen absorption that causes the break occurs in the laboratory at 1216 Å, this indicates that the five galaxies each have a significantly high redshift: approximately $z \approx 8$. These galaxies are among the most distant that have yet been identified, with a lookback time of roughly 13 billion years. They are seen as they were only 700 million years after the Big Bang, demonstrating the power of this technique.

Once the approximate redshift (or lookback time) was determined for all HDF galaxies by identifying the wavelength where the spectrum break occurred for each galaxy, a comparison of the brightness of that galaxy at UV wavelengths versus its brightness at longer wavelengths allowed a determination of the number of hot, recently formed stars in the galaxy. In this way, the Madau group built up the history of the rate at which stars have formed in galaxies over cosmic time.

Figure 11.3 shows the results of the Madau et al. determination of how the formation of stars proceeded over the past 12 billion years, which is as far back in time and distance as the HDF was able to detect galaxies. Within considerable uncertainties, it can be seen that star formation at redshift $z = 4$ was not prolific, because stars and galaxies were still in the process of forming. At a redshift of 3, which is 1/2 billion years later, the rate had increased significantly, to the point where

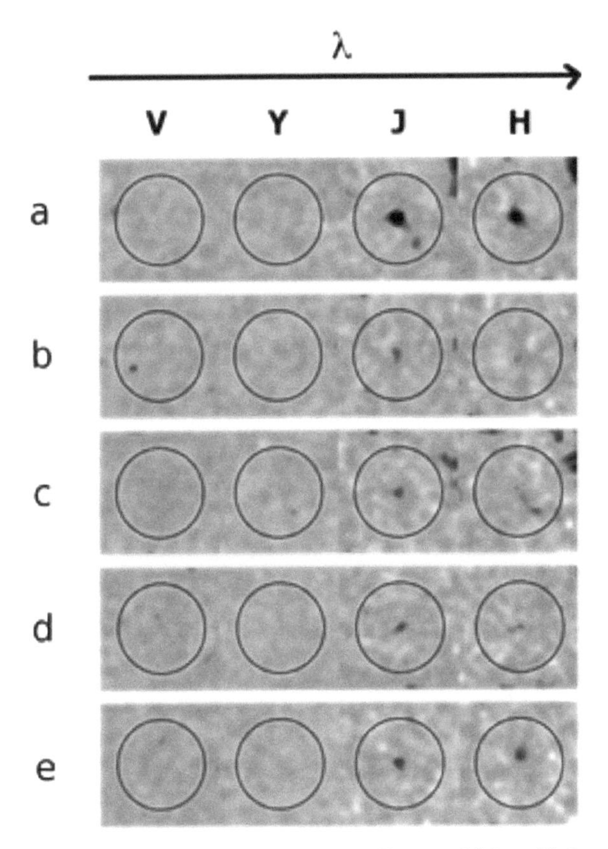

Figure 11.2. Images of five galaxies are shown in four separate filters at visible and infrared wavelengths. With no detection in the images through the shorter, visible-wavelength V and Y filters, but definite detections in the infrared J and H filters, these "drop out" galaxies must have redshifts of $z \approx 8$. (Reproduced from Trenti, M., et al., Overdensities of Y-Dropout Galaxies from the Brightest-of-Reionizing Galazies Survey: A Candidate Protocluster at Redshift z ≈ 8, ApJ, 746, 55, © 2012. The American Astronomical Society.)

it exceeds the current rate for galaxies with lower redshifts in the interval $z = 0–1$. Stars continued to form at relatively high rates until redshifts of $z = 1–2$ (7–8 billion years ago), after which time it began to drop dramatically. The interpretation of this star formation history is that the gas in the universe—that same gas as in our Milky Way galaxy, in the form of beautiful nebulae—has steadily been used up in the process of gravitationally condensing into stars. There is progressively less of it available to form new stars, signaling a slow decline in those sources of energy in the cosmos that will continue into the future. The fundamental implication of the history of star formation is clear: the gas available for star formation in the cosmos is largely used up, and therefore as stars expend their source of energy and die out, there will be no new stars to replace them. The universe is heading toward a dark future billions of years from now.

A more recent version (shown in Figure 11.4) of the star formation rate has been assembled from more data and a variety of different methods that is more precise than the original "Madau plot," as it has been called. Star formation in the universe

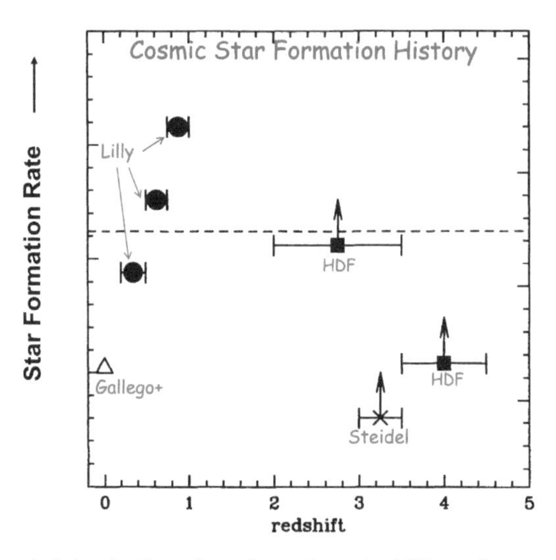

Figure 11.3. The first rough determination of star formation rates billions of years ago, calculated from the HDF for galaxies with redshifts of $z = 3$ and 4. Lower redshift data are from an earlier low-redshift survey by Lilly and collaborators. (Reproduced from Madau, P., et al. 1996, High-redshift galaxies in the Hubble Deep Field: colour selection and star formation history to z ~ 4, MNRAS, 283, 1388, by permission of Oxford University Press.)

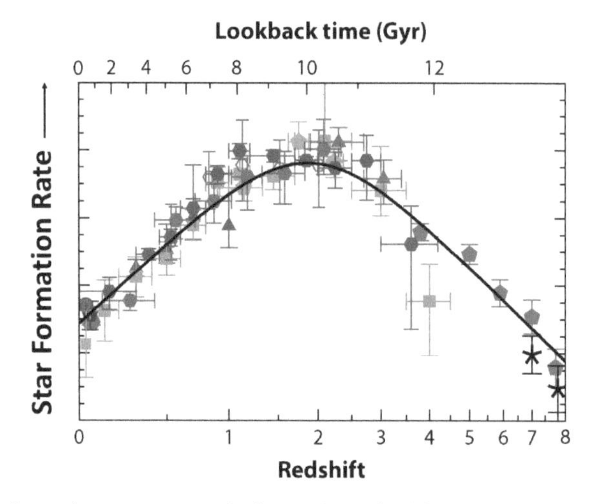

Figure 11.4. The star formation rate over cosmic time as determined from multiple studies after the HDF. Star formation initially started slowly, as galaxies first formed. It ramped up for the first three billion years after the Big Bang, but has steadily decreased to the present epoch as less gas has become available. (Reproduced with permission of Annual Review of "Cosmic Star-Formation History", Madau, P., & Dickinson, M. 2014, ARA&A, 52, 415, © by Annual Reviews, http://www.annualreviews.org.)

peaked at a time 10 billion years ago and has been declining since. The determination of the star formation rate created a huge wave of interest when the Madau group result was first published, and it remains one of the more vigorous topics of research in astrophysics. The history of star formation in the cosmos remains one of the most truly important results to have come out of the Hubble Deep Field.

One of the reasons that star formation results are known much better now than when the HDF results came out is that a new procedure has been developed that enables the redshifts of distant galaxies to be measured without requiring their spectra to be obtained. Such redshifts are called *photometric redshifts*, and they have completely transformed the way in which galaxy evolution and cosmology are now approached observationally.

Substantial amounts of time on modern telescopes, whether in space or on the ground, are scheduled for imaging and spectroscopy. Types of instruments other than cameras and spectrographs are steadily making incursions into telescope schedules, but imaging and spectroscopy remain the bedrock observations of modern observational astronomy. Imaging is easy to understand; spectroscopy less so. A picture may be worth a thousand words, but explaining a spectrum usually *requires* a thousand words. The fact remains, examining the spectrum of an object is more akin to putting DNA under a microscope than anything else an astronomer can do for objects we can only study by observing their radiation—electromagnetic or gravitational. When astronomers want to understand an object, they go to great efforts to get a good spectrum. For bright objects that are nearby, this is an easy task; for distant, very faint galaxies, it is very tedious and time-consuming to do—if can be done at all—because of their faintness.

A spectrum reveals the basic characteristics of an object, including—but not limited to—its elemental composition, temperature, density, velocity, and (in some instances) its mass. Because a spectrum consists of the light of an object spread across different wavelengths, or colors, the image of an object taken through a filter that allows only certain wavelengths of light to pass through it may be considered only one small piece of a rudimentary spectrum. Individual images of an object that are taken through separate filters that transmit different wavelengths do, in principle, constitute a crude spectrum. Some spectral information may therefore be gleaned from a series of images of objects taken in ultraviolet, blue, yellow, and red filters—which is exactly what the HDF images of the Deep Field galaxies provided.

Images of galaxies in only four filters do provide spectral details or information, but it is far too sparse to yield much information about the galaxies. However, it might just be enough to enable a valid determination of their redshifts, and therefore their distances and lookback times. The HDF team was aware of this prospect in the selection of filters to be used for the HDF exposures, but we had some doubt whether redshifts obtained in this way would be sufficiently reliable to be useful. Other investigators were also intrigued by this possibility and were motivated to work with both the HDF filtered images and the 125 Keck spectroscopic redshifts of the HDF galaxies, in order to devise ways by which the characteristics of the 125 galaxies with known redshifts might be extended to many more galaxies detected in the Deep Field.

When the Keck spectra of the 125 galaxies in the HDF were compared with the spectra of well-studied galaxies not too distant from the Milky Way in the *local* universe, it was found that the spectra of the large majority of galaxies had some similarities. In particular, galaxy spectra, some of which are shown as plots of their intensity versus wavelength in Figure 11.5, show distinguishing features that can make certain traits of the galaxy identifiable by measuring the brightness of the galaxy from images in different wavelength intervals. For example, in Figure 11.5, the aforementioned drop in the brightness of galaxies at wavelengths shortward of 4000 Å is due to hydrogen, and it is quite prominent. Furthermore, the narrow spikes in emission, called emission lines, at wavelengths near 3700 and 5000 Å are quite noticeable in some galaxies. These features come from diffuse hot gas, not stars, in those galaxies. If an image of such a galaxy were taken with a filter that included those wavelengths, the galaxy would appear notably brighter than other galaxies. More significantly, if such a galaxy with those emission lines were at a large distance, such that the wavelengths of the emission lines were redshifted to longer wavelengths, that galaxy would appear much brighter in an exposure taken with a filter that transmits light around, say 7000 Å (very red), but not around 3700 Å (very blue). In this way, extending the procedure used by the Madau team and Chuck Steidel of Caltech, astronomers could deduce the redshift of that galaxy simply by taking exposures of the galaxy through filters with transmission band passes at 3700, 5000, and 7000 Å. The redshift of the galaxy would be determined without actually having to obtain a spectrum of the galaxy. The images obtained in three or four filters of a group of galaxies—which can be done much more quickly for an entire group of galaxies, many of which may be quite faint, rather than obtaining one spectrum of just one galaxy at a time—would be adequate to determine the

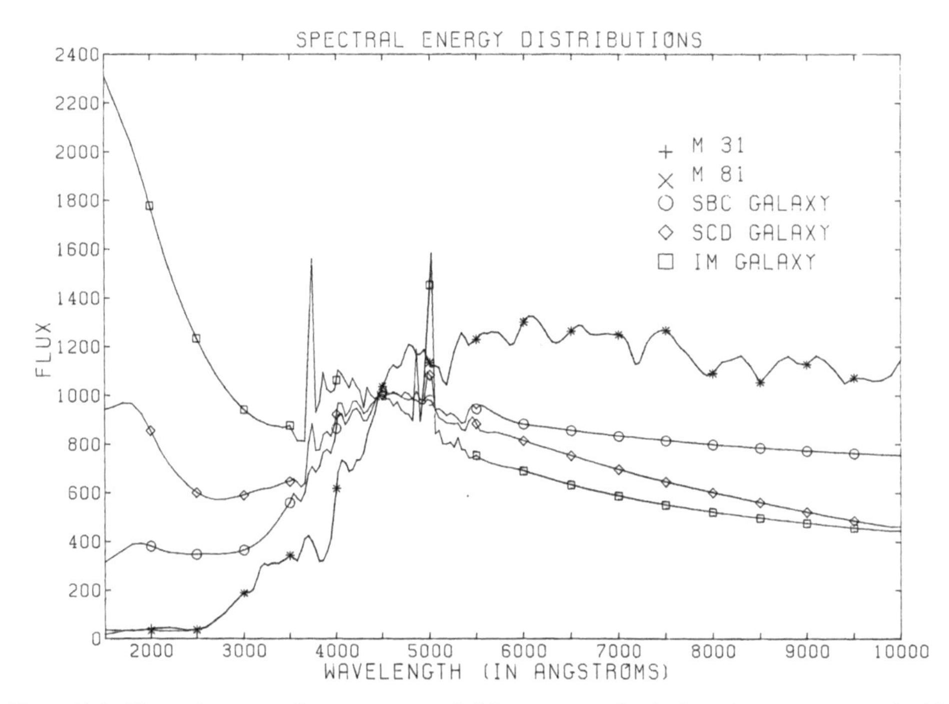

Figure 11.5. Observed spectra of an assortment of different types of galaxies. These spectra, acquired by telescopes on the ground and in space to obtain UV data, were used to predict the expected intensities of galaxies at different redshifts when imaged through different wavelength filters. (Reproduced from Coleman, G. D., Wu, C.-C., & Weedman, D. W., Colors and Magnitudes Predicted for High Redshift Galaxies, ApJS, 43, 393, © 1980. The American Astronomical Society.)

approximate redshifts of all the galaxies. Such redshifts, determined from the relative brightness in different wavelength filters from separate images or pictures of galaxies, are called *photometric redshifts*.

Photometric redshifts are straightforward to obtain for even the most distant, faint galaxies and they are made more reliable when the number of filtered images is increased. Their utility for the Deep Field was demonstrated in a key research paper by University of Toronto astronomers Sawicki, Lin, and Yee, who used the above procedure to compute initial photometric redshifts for the first 50 galaxies in the HDF that Keck 10 m observers had acquired. In comparing their photometric redshifts with the true spectroscopic redshifts from the Keck spectra, they found quite good agreement between the two, as is shown in Figure 11.6. Photometric redshifts obtained from images taken in 4–5 filters have an accuracy of around 10%, but when 7–8 filtered images are used, the accuracy improves to better than 5%. Such redshifts provide the lookback times for the galaxies, i.e., the epoch in history when they are being observed, and are key to piecing together the history of how galaxy characteristics have evolved in time since the epoch of the Big Bang.

Because the data from the HDF coupled with the Keck spectroscopic redshifts enabled the process by which reliable photometric redshifts could be obtained for many galaxies, their use as essential information in surveys of galaxies as a research topic has become the linchpin of the field. What once was so difficult to obtain for groups of faint objects is now feasible, and this has given huge impetus to

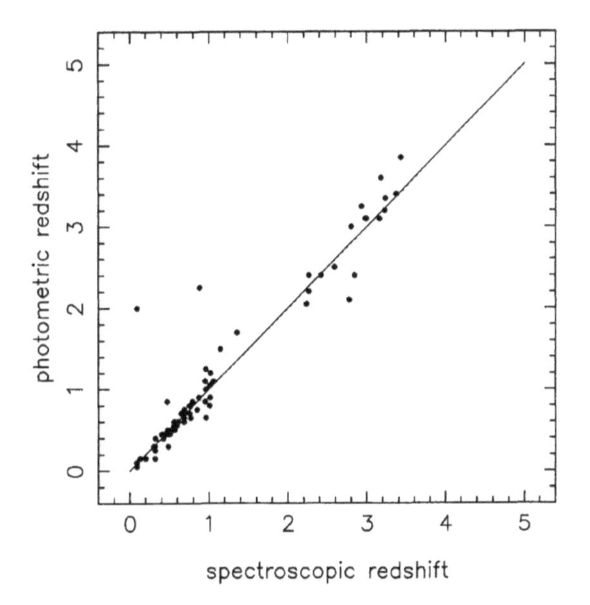

Figure 11.6. The photometric redshifts determined for 50 of the brighter galaxies in the HDF, based on their brightness (or intensity) in the four different filters, are compared with their true spectroscopic redshifts determined from their spectra acquired by the Keck telescope. A good correlation exists, with only a few outliers, indicating that the photometric redshifts are reasonably accurate. (Reproduced from Sawicki, M. J., Lin. H., & Yee, H. K. C., Evolution of the Galaxy Population Based on Photometric Redshifts in the Hubble Deep Field, AJ, 113, 1, © 1997. The American Astronomical Society.)

cosmological studies of star and galaxy formation in the early universe. Photometric redshifts of galaxies obtained simply by taking a succession of a few exposures with different wavelength filters have transformed our study of the universe by providing us a knowledge of the distances to virtually every galaxy that telescopes are able to detect in an image. This is a remarkable accomplishment. Even as a lifelong spectroscopist myself, one who believes that full understanding of every process and environment in the universe requires spectroscopy, I consider the validation of photometric redshifts that came out of the HDF to be its most important scientific accomplishment.

Hubble Deep Field and the Distant Universe

Robert Williams

Chapter 12

Synopsis and Subsequent Deep Fields

Three significant scientific results emerged from the original HDF observations carried out in 1995. First, the evolution of discrete galaxies was revealed to be largely a "bottom up" process. High-redshift galaxies started out as much smaller objects than galaxies of the present epoch. They contained a large fraction of relatively hot massive stars and had a chaotic morphology as shown so clearly in Figure 10.6. By interacting gravitationally with neighboring star groups and gas clouds, they steadily increased in size and complexity via mergers to develop into the large, more ordered and symmetric galactic systems that occupy the universe today. Second, the rate at which stars formed in young galaxies a few billion years after the Big Bang was much greater than the rate at which stars currently form in galaxies in the present epoch. Much of the gas out of which stars form has been used up in the past 10–13 billion years, and although many stars do eject substantial mass into space when they explode as supernovae, thereby partially replenishing the gaseous reservoir, it is not enough to replace the gas previously used up in the formation of the early generations of stars. The universe is indeed slowly dying out from the depletion of gas that can form into stars.

Third, the 125 spectra and redshifts of HDF galaxies from the Keck 10 m telescope were crucial in augmenting the Deep Field image so a process of determining photometric redshifts could be implemented for galaxies. Along with the HDF images taken in separate wavelength bands, those spectra helped create a fundamental paradigm shift in cosmology—fairly reliable distances can henceforth be obtained for almost any galaxy without having to go through the time-consuming process of acquiring its spectrum. Instead, one may simply obtain images of objects in filters that transmit different wavelengths. Taken together, these three results from the HDF substantially advanced the study of galaxy formation and evolution.

Apart from the aforementioned significant scientific results from the HDF, the campaign also had an impact in changing the culture normally associated with large observing programs on telescopes, including the data distribution. The HDF was

doi:10.1088/978-0-7503-1756-6ch12

undertaken separately from the normal peer review process. It would very likely not have happened had the HDF Project been proposed as part of the usual telescope allocation procedure. It was too risky and it required much too much time on a highly sought-after resource like the *Hubble Space Telescope*. After the data were received, the team reduced the huge dataset in seventeen days and made them immediately available to everyone. The HDF team waived its right to keep the data to itself for one year, demonstrating that doing research with one's own proprietary data while allowing the community full and equal access to them does benefit the overall advancement of knowledge.

We have confined our focus to the central objective of the initial Hubble Deep Field campaign, following the thread of what first motivated it, how the process of implementing it evolved over time, and what we learned from it. While the primary HDF was being planned, additional features of the campaign were thought of that we had not conceived of initially, so they were incorporated into the observations and data analysis. New ideas continued to emerge after the HDF data were released, such that a strong interest developed within the research community to undertake additional deep fields with the *HST* that could extend the discoveries from the HDF image. All of these factors combined to make the HDF the beginning of what was to become a series of deep probes into the past that have increased our understanding of how the universe came to its present form following the Big Bang.

The *HST* is capable of operating multiple instruments simultaneously, so adjacent areas of the sky are often observed at the same time with two different instruments. The location of the entrance apertures of all the instruments are near each other at the back end of the telescope, so areas of the sky that are observed simultaneously are all within less than half a degree of each other. The HDF team realized we could operate the NICMOS instrument developed by Dr. Rodger Thompson and colleagues at the University of Arizona, which is sensitive to infrared wavelengths, at the same time we were taking exposures of the primary HDF field in visible wavelengths with the WFPC2 camera. The NICMOS detector was smaller and less sensitive than the WFPC2 mosaic of detectors, so the NICMOS deep field did not achieve the coverage, depth, and resolution of the WFPC2 visible HDF, but it did produce its own deep image that characterized distant galaxies in the infrared.

Some months after the HDF had been completed, it was already clear that the dataset represented a unique window into the past that should be exploited further. There were a goodly number of colleagues who approached us and acknowledged the beneficial research results that the HDF image produced, but there was also a question as to whether the HDF had truly imaged a representative part of the cosmos. Was our "undistinguished" patch of sky really so? Being a "one off," our core sample of the universe might by chance have sampled an uncharacteristic volume of the cosmos, which has voids, clusters of energetic galaxies, and regions around quasars that might not be representative of the universe. There was only one way to find out: repeat the HDF campaign in a completely different location in the sky, far removed from the direction of the original field.

So, repeat the original HDF we did, three years later and using the same procedures as before, augmenting the original team with a number of additional scientists from the Institute. The successor effort was carried out in the Southern Hemisphere, called the Hubble Deep Field-South (HDF-S), and it showed a basic structure that was similar to that of the original HDF. Unlike the northern field, however, the southern field contained a number of very red, elliptical-shaped galaxies, thus demonstrating a degree of the variance that could be expected among different regions of the universe.

An impressive series of deep field surveys carried out on the *Hubble* have followed the original HDF over the past 20 years, to improve on what the HDF undertook. A major reason for the success of the subsequent surveys was the installation of improved instruments and detectors on the *HST* from each of the servicing missions. Within a few years of the HDF Project, the capabilities of the WFPC2 camera had been surpassed by newer instruments named ACS, STIS, and WFC3 that provided for much improved sensitivity, smaller pixel size (which provided better resolution), broader wavelength coverage that enabled observations at both shorter and longer wavelengths into the ultraviolet and infrared portions of the spectrum, and large detector sizes that imaged larger areas of the sky. In every regard, the newer instruments enabled the next generation of deep fields to penetrate further back in time over larger areas of the sky and cover a much wider wavelength region of the spectrum in less time than had been possible in late 1995 with WFPC2. These successive post-HDF deep fields represent significant improvements over our original observations because they have detected more distant galaxies closer to the time of the Big Bang, as well as smaller, fainter galaxies earlier in their evolutionary development.

The chain of successive deep fields with the *Hubble* has progressively revealed the changing nature of the universe as we look back in time. Some of the follow-on observational campaigns have been made with facilities other than the *HST*, and an important component of their success has been the fact that many of them have observed the same region of sky, enabling those observations to build upon the unique observations that previous deep fields made of that same region of sky. For example, the original HDF field was subsequently imaged not just in the infrared with the *Hubble* NICMOS instrument, but subsequently with the infrared-sensitive satellite *Spitzer Space Telescope*. It was also imaged in the X-ray region with the *Chandra* X-ray Observatory and additional X-ray satellites that were developed and operated by the European Space Agency.

We provide below a brief summary of some of the major deep field campaigns undertaken with the *Hubble* that followed the original HDF (1995) and HDF-S (1998) campaigns. In our description of the surveys, we refer to them by their more convenient acronyms.

GOODS (2002): The Great Observatories Origins Deep Survey began the next generation of deep fields to follow the HDF on *Hubble*; it was driven by an interesting confluence of events. One of these was the installation of the new Advanced Camera for Surveys (ACS) instrument on the *HST* in 2002. This camera had a much larger and more sensitive detector than WFPC2. Secondly, NASA was soon to launch the infrared SIRTF satellite, subsequently re-named *Spitzer Space*

Telescope, which would survey the sky for sources of infrared radiation, some of which would certainly be interesting galaxies. Regions of star formation are commonly associated with concentrations of dust that cause them to be bright in the IR, and the modest size of the *Spitzer* meant that it had low spatial resolution. Interpretation of *Spitzer* observations would benefit from IR objects being imaged in the visible by the *HST*, so the two satellite telescopes could work in tandem to image the same areas of the sky to deep levels.

The above factors notwithstanding, a significant motivation for GOODS was the important "after-the-fact" discovery of a very distant supernova in the HDF in 1999 by Ron Gilliland, Mark Phillips, and Peter Nugent. At the time of the HDF, there were independent teams trying to determine if changes existed in the expansion rate of the universe, i.e., if the Hubble constant was not really constant in time. It was believed that the gravitational attraction of galaxies could be slowing cosmic expansion. Because the effect might be extremely small and difficult to measure, it was essential to use distant luminous objects, such as supernovae, to make the measurements of their recession velocities, as these would yield the cosmic rate of expansion, i.e., the Hubble constant, at the time long ago in the past at which they were being observed.

The discovery of the distant HDF supernova in a galaxy whose redshift, and therefore distance, had been measured had a pronounced effect on cosmological studies. Research teams immediately realized that the best way to determine whether forces were causing a deceleration or acceleration in the universal expansion rate, measured initially by Edwin Hubble in the 1920s, was to try to detect more distant supernovae. The best way to accomplish this was to make observations in sequence at different times so the change in brightness of the supernovae would draw attention to their existence and their distances could be determined. The GOODS team, headed by original HDF team member Mauro Giavalisco, therefore crafted their survey to take many exposures, but instead of taking them consecutively in one 2–3 week interval as we had done for the HDF, they spread the exposures over weeks and months so the presence of supernovae in the images would become apparent.

An important aspect of the GOODS observations was resolving the quandary regarding the width versus the depth of the observations. This dilemma has dogged all of the deep field surveys. On one hand, one wants to observe the faintest possible galaxies, i.e., go deep. On the other hand, one also wants to include a sufficiently broad area of the sky to include a substantial number of galaxies. For a fixed amount of observing time available on the *HST*, each deep field survey has had to decide whether to expose as long as possible with one telescope pointing or to observe less time per pointing so that adjacent areas of sky can be imaged and mosaicked together to form one larger, albeit shallower, field.

Compromises are necessary, and they have led to the adoption of a "wedding cake" structure for most of the deep *HST* fields. Typically, one small patch of sky from a single *HST* pointing forms the deep core image, which is then surrounded by a ring of shorter, shallower exposure images. The original HDF actually incorporated this feature by taking one-orbit exposures of "flanking fields" that surrounded the HDF field. The ensemble of images form a wedding cake structure in terms of faintness detection limit versus amount of sky coverage. The multiple-epoch images of the GOODS survey and the relatively large area of sky covered were important facets of their deep fields in the northern and southern hemispheres of the sky that did, in fact, detect a number of distant supernovae. Those supernovae were studied by independent teams measuring the cosmic expansion rate and they became key elements in the determination of what turned out to be, quite unexpectedly, cosmic *acceleration* rather than *deceleration*. The deep fields of the *HST* thereby played important roles in the determination of *dark energy*, an important discovery that was honored with the 2011 Nobel Prize in Physics.

Ultra Deep Field (2004): Following the GOODS project, Steve Beckwith, Institute Director from 1998–2005, conducted a study of possible observing programs that would benefit from the assignment of many orbits of DD time. Given that the GOODS campaign had sacrificed some depth in faintness detection limits in order to broaden sky coverage, Beckwith concluded that a new *HST* deep field would return the best science if it were carried out with the same Advanced Camera for Surveys (ACS) instrument, but to a much deeper detection level in a narrower region of sky and also in an important, longer-wavelength filter band that would detect higher redshift galaxies. The ACS had significantly better sensitivity and a larger detector than the WFPC2 camera, so a larger area of the sky could be covered. Drs. Beckwith and Massimo Stiavelli moved forward with the Ultra Deep Field (UDF) project, which used 400 *HST* orbits in late 2003 to push the limits of *HST* detection significantly beyond the earlier deep fields. The UDF succeeded admirably in producing what remains one of the most enduring images of the cosmos at visible wavelengths, shown in Figure 12.1.

In the years following the release of the UDF, it served as a basis for subsequent deeper images of the same field by teams of researchers who augmented the original UDF image with additional data obtained from a variety of other facilities. The extension of the UDF has taken several forms, including the addition of data from a broad range of wavelength bands and the acquisition of spectra of as many UDF objects as possible. The additional wavelength data have come from radio telescopes on the ground and satellite telescopes that are sensitive to a broad range of wavelengths in the X-ray, ultraviolet, and infrared spectral regions. Additional images and spectra have come from large ground-based telescopes and also from a more recent replacement instrument on the *HST*, the Wide Field Camera 3 (WFC3), which has unmatched sensitivity in the infrared. Combined into one dataset, the additional data have given astronomers our deepest views of the universe in the optical and IR wavelengths

The extension of the UDF imaging has been spearheaded by several groups of astronomers whose teams have maintained a healthy competition. Drs. Richard Ellis (Caltech and University College London) and Garth Illingworth (Univ. of California/Santa Cruz), who were both members of the original Institute Advisory Panel that debated deep field strategy in 1995 March as a prelude to the HDF, each formed separate research teams and developed new techniques that have yielded

Figure 12.1. The Ultra Deep Field, a pencil-beam image of the cosmos that shows faint galaxies out to distances greater than 12 billion light years. Taken with the ACS instrument on the *HST*, it imaged a larger field to a fainter level than the original HDF [credit: S. Beckwith/UDF Team]. A larger version of this figure is available at http://iopscience.iop.org/book/978-0-7503-1756-6.

excellent research results in the two decades since the original HDF. An enhanced UDF image was compiled by the Illingworth group in 2013, showing 2963 separate images that they labelled the eXtreme Deep Field (XDF), which they announced to be the "deepest image of the sky ever taken." In the same year, the Ellis team produced a comparable enhancement of the UDF by augmenting it with long exposures of that region of sky in IR wavelengths obtained with the WFC3 instrument on the *Hubble*. Because cosmic expansion shifts the luminous radiation from distant galaxies into the IR, infrared wavelengths are indeed optimal for the study of the distant universe. The Ellis group called their image UDF12, and also pronounced it to be "the deepest ever image of the sky." There is truth to both claims, as each team successfully obtained the spectra of their faintest galaxies in different ways to reveal galaxies at distances approaching 13 billion light years. This distance corresponds to redshifts $z > 9$, remarkably close to the epoch of the Big Bang and the initial formation of large structure in the cosmos. Recall that the original HDFs taken with WFPC2 were able to detect galaxies only to a distance of 12 billion light years, corresponding to redshifts of $z = 4$.

CANDELS (2010–13) + Hubble Frontier Fields (2014–16): Enhancements of the Ultra Deep Field have only been one part of the progress of imaging the sky to faint levels. Much attention has been paid to the deep campaigns whose images capture the most distant galaxies, but other *Hubble* surveys emphasizing the wide sky coverage aspect of deep fields have also produced groundbreaking results. The COSMOS and AEGIS survey fields, which were peer-reviewed projects involving large teams that were awarded *HST* time in 2005, are examples of imaging of relatively wide swaths of the sky. The COSMOS images made use of the fact that, as proposed in Einstein's Theory of General Relativity, mass does warp space–time in a way that causes it to act as a lens in bending light rays. By studying the detailed forms of galaxies whose shapes have been distorted by the presence of large concentrations of mass along the sight line from Earth to the observed galaxies, the total amount of gas required to produce the observed warped shapes was determined. The mass far exceeded the amount of mass needed to produce the observed light from the region, thereby demonstrating the existence of large concentrations of matter that do not interact with radiation—a new type of matter with properties unfamiliar to scientists. The ability of the *Hubble* to detect many faint galaxies whose shapes have been distorted by the presence of intervening *dark matter* has enabled it to be mapped. Remarkably, the location of dark matter has been found to be correlated with the presence of normal *baryonic* matter. The mapping of dark matter constitutes one of the important results of the *HST's* deep fields.

Two very recent deep field surveys, which have consumed significant amounts of observing time on the *Hubble*, merit our attention: they may represent the final deep field programs to be undertaken by the telescope before it starts making observations collaboratively with its larger successor telescope, the *James Webb Space Telescope* (*JWST*). The CANDELS survey was awarded an impressive 902 orbits of *Hubble* time to take exposures of five regions of the sky that had previously been imaged with other instruments and filters as deep fields. One important feature of this survey, which it shared in common with the COSMOS and AEGIS fields, was the commitment of an extensive amount of observing on the Hawaii Keck 10 m telescope with a new technology spectrograph to get spectra of thousands of the faint galaxies observed in the *Hubble* images. These spectra provided accurate distances and lookback times for the galaxies and also information on the characteristics of the stars within the galaxies, as well as their motions. The velocity information for the stars might not seem to be of importance for cosmology, but it is, in fact, an essential indicator of many fundamental properties of activity in a galaxy. The CANDELS team was able to infer rates of star formation, the production of elements, the evolution of the stars, and the history of mergers of the galaxies from their spectra. The important focus on incorporating spectral information in the analysis of deep field images owes much to the dedicated efforts of Dr. Sandra

Faber. She and Harry Ferguson led the CANDELS campaign to develop innovative spectrographs for the Keck Observatory telescopes so they could be applied to understanding the role of the stars and gas in driving the evolution of galaxies.

The most recent—and possibly final—*HST* deep field survey is a significant departure from all previous *Hubble* deep field campaigns, in that it uses an ingenious method to enhance detection of distant galaxies via gravitational lensing. Instead of directly imaging what are believed to be typical, or undistinguished, regions of the sky, the Hubble Frontier Fields survey, led by Institute Director Matt Mountain and Dr. Jennifer Lotz, selected six rich clusters of galaxies to image: these clusters' large masses serve to gravitationally focus the light from more distant galaxies. The gravitational lensing by massive galaxy clusters along the line of sight to more distant galaxies behind the clusters serves to amplify the brightness of the distant galaxies. The lensing produces beautiful arcs, like those shown in Figure 12.2, that are distorted, magnified images of more distant galaxies that may be too faint to be detected were they not lensed by the foreground cluster of galaxies.

The Frontier Fields observations produced a series of clear images of the galaxy clusters, around which very faint galaxies can be seen. Many of those galaxies are images of more distant background galaxies that have been rendered visible due to lensing by the massive cluster. What is especially important is that the images of lensed background galaxies tend to be focused around the lensing cluster in different positions that depend on their distances from us. Astronomers have used *Hubble* images to map out the regions around the lensing clusters where galaxies having a certain distance, or redshift, might be located, an example of which is shown in Figure 12.3. A red line *critical curve* has been drawn around one such cluster, Abell 2744, that indicates where galaxies having a distance of 13 billion light years, corresponding to a redshift $z = 11$, are expected to be located. Studies of the Frontier Field clusters are still underway, and a number of very faint, lensed galaxies have been found around the lensing clusters in positions where their suspected distance is about 13 billion light years. Their true distances must be determined by obtaining their spectra, but if confirmed, they represent galaxies that exist only 700 million years after the Big Bang—a long time span by human standards but short by galaxy evolution and cosmological standards. The Frontier Fields are providing us with perhaps the best method to identify the earliest galaxies to have formed after the Big Bang.

With each *Hubble* servicing mission, new instruments with better sensitivity were installed on the telescope, increasing the capability of the *HST* to probe the cosmos. The significance of observing deeper into epochs closer to the time of the Big Bang has retained an allure that is difficult to resist. The annual allotment of Director's Discretionary time on *HST* has continued to be used as an inducement to support programs that require a larger number of orbits than are likely to be approved by the normal peer review process. There are many areas of astrophysics for which large amounts of *HST* time are needed to provide first-rate results, but few of them absolutely demand the extremely long exposures on a single field that are unique to observations devoted to cosmological studies.

At the time of this writing, the *Hubble Space Telescope* has been in operation for 28 years and is still working flawlessly even though its most recent servicing mission

Figure 12.2. Galaxy cluster J1038+4849, taken with the *HST*, showing beautiful arcs that are the distorted images of very distant galaxies behind the two bright foreground galaxies whose masses are producing the distortion. The blue arc is that of a galaxy with many young, hot stars; the red arc represents a galaxy dominated by older, cooler stars [credit: NASA/ESA].

was now nine years ago, an interval twice as long as any that occurred between servicing missions when the Space Shuttle was still operational. The *HST* will not continue to operate forever, yet it is still providing the most detailed views of objects near and far in the universe. Just months ago, its sharp vision gave astronomers a detection in visible wavelengths of the burst of radiation that was produced by the very distant pair of merging neutron stars that produced the gravitational waves detected recently by the LIGO and VIRGO gravitational wave detectors.

The *HST*'s successor telescope, the *JWST*, is scheduled for launch in the coming years. It has plans to undertake its own deep field surveys that will extend astronomers' vision even further into the early universe. The *JWST's* sensitivity to infrared wavelengths, where distant galaxies radiate the bulk of their light due to

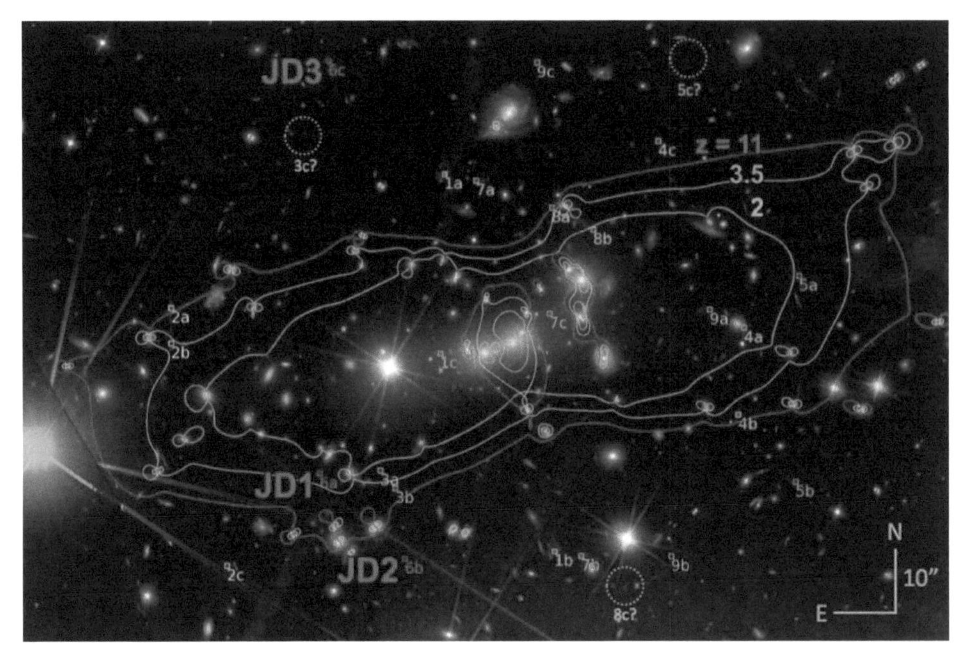

Figure 12.3. The lensing cluster of galaxies MACS0647+70 was a candidate cluster for the Hubble Frontier Fields study. The colored contours represent regions where gravitationally lensed images of distant galaxies beyond the cluster should occur having redshifts $z = 2$, 3.5, and 11, corresponding to distances of 8, 10, and 12 billion light years. The faint images labeled JD1, JD2, and JD3 are multiple images of a distant, lensed galaxy candidate having $z \sim 11$, representing one of the most distant galaxies yet observed (Reproduced from Coe, D., et al, Clash: Three Strongly Lensed Images of a Candidate z ≈ 11 Galaxy, ApJ, 762, 32, © 2013. The American Astronomical Society.). A larger version of this figure is available at http://iopscience.iop.org/book/978-0-7503-1756-6.

the cosmic expansion that produces their high redshift, will certainly enable the detection of objects closer to the time that cosmic gas clouds first began to assemble into stars and galaxies. Those discoveries will no doubt be just as transformational as were the *Hubble's*, and they await the future to be revealed and understood.

The *Hubble's* deep fields generated a surge of interest among the scientific community and public because of its unmatched ability to peer close to the fog that surrounds the epoch of creation. No one wants the tide of *Hubble* discoveries to ebb, but end it will, eventually. The more we understand the cosmos, the more it strengthens our connection to it. Our DNA is stamped with the chemistry that took place in the gas clouds that emerged from the Big Bang. The link between amorphous objects that began as microfluctuations in the Big Bang to form stars and galaxies billions of years ago to conscious life with all its complexity becomes more meaningful with time and successive discoveries. This link has profound significance for a humanity that so strongly needs the cosmic imprint of our common brotherhood stamped on us. For those of us who have been able to commandeer a forefront facility like the *Hubble Space Telescope* to satisfy our own curiosity about a universe that displays such power and splendor, studying the cosmos has no equals.

CPSIA information can be obtained
at www.ICGtesting.com
Printed in the USA
BVHW022104070119
537252BV00001B/1/P

* 9 7 8 0 7 5 0 3 1 7 5 4 2 *